The
Lipoprotein Molecule

NATO ADVANCED STUDY INSTITUTES SERIES

A series of edited volumes comprising multifaceted studies of contemporary scientific issues by some of the best scientific minds in the world, assembled in cooperation with NATO Scientific Affairs Division.

Series A: Life Sciences

Recent Volumes in this Series

The series is published by an international board of publishers in conjunction with NATO Scientific Affairs Division

A	Life Sciences	Plenum Publishing Corporation
B	Physics	New York and London
C	Mathematical and Physical Sciences	D. Reidel Publishing Company Dordrecht and Boston
D	Behavioral and Social Sciences	Sijthoff International Publishing Company Leiden
E	Applied Sciences	Noordhoff International Publishing Leiden

The Lipoprotein Molecule

Edited by
Hubert Peeters
Lipid and Protein Department LBS
Brussels, Belgium

PLENUM PRESS • NEW YORK AND LONDON
Published in cooperation with NATO Scientific Affairs Division

Library of Congress Cataloging in Publication Data

Nato Advanced Study Institute on the Lipoprotein Molecule, Bruges, Belgium, 1977
  The lipoprotein molecule.

  (NATO advanced study institutes series: Series A, Life sciences; v. 15)
  Includes index.
  1. Blood lipoproteins—Congresses. I. Peeters, Hubert, 1919-   II. Title. III.
Series.
QP99.3.L52N37 1977   612'.0157   78-2388
ISBN 0-306-35615-5

Proceedings of the NATO Advanced Study Institute on the Lipoprotein
Molecule held in Bruges, Belgium, May 8—20, 1977

© 1978 Plenum Press, New York
A Division of Plenum Publishing Corporation
227 West 17th Street, New York, N.Y. 10011

Printed in the United States of America

Preface

The amount of recent information collected about the molecular composition, structure, and function of the plasma lipoproteins, in man as well as in experimental animals, is very large. In this volume an assessment of our knowledge as it stands today has been organized within a framework of four general topics, the first about analytical data, the second about structure, the third about metabolism, and the last about molecular variation and pathology. Thus the analytical, theoretical, experimental, and applied aspects of the topic have been treated in conjunction with each other.

The analytical data in human and nonhuman primates were obtained after ultracentrifugal or electrophoretic separation of the protein class from the native serum. Comparisons of various methods were not forgotten. The main categories of lipoproteins, namely High Density (HDL), Low Density (LDL), and Very Low Density (VLDL), are treated separately, and lipid moiety and the polypeptide moiety are carefully described.

In the theoretical field of reassembly of lipoprotein particles by relipidizing delipidated polypeptides, the structural aspects of lipoprotein and the stabilizing effect of phospholipids on the polypeptide structure were studied. The opposite direction of the process, namely degradation of a lipoprotein by means of lipolytic enzymes, has also been investigated and described.

Some inborn errors of metabolism and some acquired disorders such as jaundice and also the pathology of atherosclerosis have been related to peculiar and often specific lipoprotein patterns. These in turn can be modified by diet or drugs, and the efforts to restore normality are largely based on the analytical and structural knowledge of lipoproteins described in the earlier sections of the course.

Discussions, round tables, and short spontaneous sessions organized by participants themselves during the course proved to teachers and students alike that this initiative was successful.

We are grateful to the faculty members for preparing their manuscripts carefully and also to Plenum Press for the speedy publication of this volume.

Hubert Peeters

Contents

CONCLUSION

Introductory Remarks

THE CHALLENGE OF THE LIPOPROTEIN MOLECULE

Hubert Peeters

Lipid and Protein Department, LBS

B-1180 Brussels, Belgium

The primaeval outline of a lipoprotein appeared in the course of 1926 and 27 when Macheboeuf presented a few short papers on the phospholipids of plasmaproteins. In 1928 he defended his doctoral thesis: Recherches sur les stérols, les lipides et les protéides du sérum et du plasma sanguin", Thèse pour le Doctorat ès Sciences, Paris. This NATO Advanced Study Institute thus coïncides with the 50th anniversary of the fundamental work from which the modern concept of lipoproteins emerged.

As a starting point he used his observation that lipids – not soluble in water – were solubilized in plasma through some of the plasma proteins. For these lipid-binding proteins he coined a new terminology – no longer in use today – le "cénapse lipoprotéique". This lipid-protein bond was not easy to disrupt as ether was unable to extract all lipids from the serum. These observations were done on horse serum which was readily available at the Pasteur Institute.

He counterchecked his hypothesis of the existence of peculiar lipid-protein structures in serum or plasma, his so called "cénapses", by mixing protein with fatty acids in the presence of bile salts, thus reconstituting some sort of a plasma lipoprotein. In 1931 he applied his views to lipoid-nephrosis – the nephrotic syndrome – because he knew, about the abnormal amounts of fat present in the plasma of these patients.

Thus the work of Macheboeuf contains the fundamental contribution to the lipoprotein concept which is the topic of this course. How did this first image evolve?

Density was applied as a separating factor for the intricate mixture of proteins which were suspected in plasma and the intriguing problem of the X protein was solved by the isolation of lipoproteins by the ultracentrifugal flotation procedure. The presence of the Donner Laboratories at this meeting shows the sustained interest for centrifugation over the bias of microequipment and the importance of the density parameter in the establishment of lipoprotein density classes.

Free zone electrophoretic patterns were seriously disturbed by lipoprotein turbidity, but on paper special lipid staining techniques helped us to relate the ultracentrifugal to the mobility patterns. In due turn immunology played its part in distributing the apopolypeptides into lipoprotein families and we are awaiting better reagents and methodology, in order to apply this knowledge on a practical clinical scale.

A morphologic approach through electronmicroscopy confirmed the ultracentrifugal concept of lipoprotein classes at the molecular or rather at the particle scale and also helped to detect anomalies connected with enzyme deficiencies. Some intriguing physicochemical problems related to the appearance of discshaped and spherical particles are being solved by means of elegant physical methods such as X-Ray scattering, NMR and several others.

However not one of these techniques has yet solved the area of uncertainty which floats around the identity of a lipoprotein molecule. It is not clear, neither from the density-, morphology-, mobility-, nor immunology parameters, that lipoproteins are precisely built. On the contrary, an aura of vagueness always prevails. The same polypeptides are present in almost every lipoprotein although in a different amount and proportions.

The chemical composition of the lipid moieties became a fruitful approach towards the identity of a lipoprotein. Once density and mobility were related to the lipid load, the distribution of the lipid classes with their many variants as well in their alcoholic as in their acyl moieties - either ester or ether - became a clue to the identity, function and metabolism of the molecule or particle.

Despite the analytical imprecision of the methods in use in lipid analysis and partly due to the problems related with the isolation of precisely defined lipoproteins, an increasing amount of data is appearing which tends to stress the rigour of the lipoprotein lipid load. Could one expect that lipids would be carried as bulk material on the delicate spine of a polypeptide? Ongoing research points towards the precise location of precise phospholipids on the polypeptide backbone. The precision of the lipid load probably reaches as far as the double bonds. It is not clear today to what extent the lipid load is determined by the polypeptide end of

the molecule and how far lipids are allowed to interexchange inside
and between their own subgroups. A question of this order is the
distribution of a given load of unsaturated double bonds over the
fatty acids of a lipoprotein molecule. We tend to believe that the
number of double bonds could partly, yet imperfectly, substitute
for each other to a certain extent, e.g. in such a way that 20
double unsaturated fatty acids could be replaced by 40 monounsatu-
rated ones. If such type of reasoning were correct the amount of
lipid in a partcile would automatically and of necessity be greater
when more unsaturated fatty acids are present.

There are more of these questions. Are permutations, inside
the lipid load, allowed all the way through or are there only some
of the double bonds available for such an exchange inside a given
lipid class? How much variation between phospholipid families is
tolerated within the same molecule? Which and how much polypeptides are required if a different lipid load is to be carried? Comparative biochemistry may bring a clue to some of these problems
and should fit in with the puzzle of human pathology. Genetic disorders in man could also help to break the code in this field.

An aspect of lipoproteins which helps to sustain their fasci-
nation is their ability to pick up lipid soluble dyes which are
readily observed on the colourless holoprotein background. Let us
forget for a moment the quantitation problems of our staining procedures and extend the idea of stainability to any chemical one can
introduce into the lipid moiety. Recently fluorescent markers have
been introduced and show to behave differently in a stoggy cholesterol milieu and more loosely in a triglyceride environment. The depolarization of fluorescence is a physical parameter to be measured
on intact and native lipoproteins without requiring their isolation
from a mixture. Such measurements are an easy means of quality control of material under separation, under delipidation or reassembly.
Along a different line this method opens the way for the study of
membrane lipoproteins which might become a topic for a later course.

Evidently the metabolic aspects of lipoproteins have been extensively studied. Ever since chyle was observed in antiquity and
used as an omen in the prediction of the future after the sacrifice
of poultry in honour of the ancestors, some relation between lipid
load and food intake had been established. The best separation
methods and refined biochemical systems are required to separate
the enzyme systems responsible for lipid turnover across the lipoprotein vehicle. Progress in the clinical field is largely based
on a modern reading of lipid disorders of the plasma of patients
and rather reminiscent of the work of the augures in antiquity.
Practical enzymatic procedures for lipolytic enzymes are still to
be further developed and standardized.

As a result of our insufficient knowledge the drug and diet

therapy of lipoprotein disorders is still in its infancy. We can expect a breakthrough only when the basic research will have clarified the position of plasma lipoproteins as a parameter of arterial pollution.

In this introduction we have been avoiding names, figures, graphs and tables. Thus those who are to follow can use the weapons and the bullets of their choice. It is the role of a general to decide on the line of attack, to draw a map, to describe the position, but to avoid the details.

If the faculty has been selected for its competence the course has been lined out for the sake of the students. If lipoproteins indeed cause turbidity in our biological fluids, they should generate clear physiological concepts in our minds and this was the purpose of this Advanced Nato Study Institute and is the aim of these proceedings.

Isolation and Composition of the
Lipoprotein Molecule

THE PROGNOSIS FOR QUANTITATIVE LIPOPROTEIN ELECTROPHORESIS

Frank T. Lindgren

Donner Laboratory, Lawrence Berkeley Laboratory

University of California, Berkeley, California

LIPOPROTEIN vs. CHOLESTEROL AND TRIGLYCERIDE DETERMINATIONS

Surely any system for classifying lipid and lipoprotein disorders will have limitations, and this is certainly true for the Fredrickson, Levy and Lees "typing" system (1). Although during the past decade this "typing" system has allowed classification of most lipoprotein disorders into six types, there has been some dissatisfaction and disenchantment in its usage. Sometime ago, two editorials (2, 3) have advocated discontinuance of "routine electrophoresis" in clinical testing. Unfortunately, this would potentially have the effect of removing a cornerstone of the typing system, namely, the direct estimation or quantification of the major lipoprotein classes.

The real limitations of the typing system as practiced by clinical laboratories were twofold: a) it failed to categorize lipoprotein abnormalities on a genetic basis, and b) it did not provide quantitative lipoprotein data. It is ironic that at just the time when the earlier crude techniques of paper electrophoresis are being replaced by refined techniques capable of yielding such quantitative data (4, 5) some investigators are recommending wholesale abandonment of electrophoresis.

One author (2) advocates discontinuance of routine lipoprotein electrophoresis, but at the same time wishes "to try to better understand lipoprotein metabolism". This goal seems incompatible with the elimination of electrophoresis, since it would appear necessary to study both qualitatively and quantitatively all of the lipoprotein classes rather than just plasma triglyceride (TG) and cholesterol (TC) levels. The author further justifies his position

9

by saying, "There is virtually a total overlap of information between plasma levels of cholesterol-triglycerides and lipoprotein
electrophoresis (exceptions: triglyceride > 350 mg/100 ml)".   I believe this is an unfortunate overstatement. The assertion that
"one could quite accurately predict the lipoprotein pattern" knowing the plasma TG and TC as well as the "appearance" of the fasting plasma is probably correct in most cases, particularly if one
is an experienced clinician. Yet, there are important exceptions
in both normal and hyperlipoproteinemic sera. A complication is
introduced by the fact that there is a similar ratio of TG/TC in
both low density lipoproteins (LDL) and high density lipoproteins
(HDL). The former molecules are atherogenic while there is evidence to suggest the latter are potentially helpful for necessary
cholesterol clearance from tissues (6). Indeed, there is evidence
to suggest that HDL is inversely related to CHD prevalence and that
this relationship is largely independent of total cholesterol and
LDL levels (7). An elevation of either 100 mg/100 ml LDL or 155
mg/100 ml HDL would raise both the serum cholesterol by 33 mg/100
ml. In such a case, it would be impossible to distinguish by lipid
values alone, whether this elevation was due to LDL or HDL. However, this could be determined by internally standardized quantitative agarose gel electrophoresis.

The implications are twofold. At a relatively moderate plasma cholesterol level (e.g., 210-240 mg/100 ml) a physician might
overlook a patient (false normal) with an elevated LDL if the HDL
is abnormally low. Conversely, with a somewhat elevated plasma
cholesterol, say 240-280, a patient with an unusually high HDL
might be unnecessarily subjected to diet and/or a drug such as cholestyramine, with no apparent therapeutic effect. Perhaps as many
as 10% of individuals screened would fall in the above categories
and thus be misdiagnosed.

Another more important example of the usefulness of electrophoresis would be the discrimination of most type II b's, III's,
and some IV's, one from another. This might involve some 30% of
all patients with lipoprotein abnormalities (8). Except for the
type III, which in addition to electrophoresis would require ultracentrifugation for confirmation, quantitative electrophoresis would
permit a more refined definition of the patient's lipoprotein abnormality. Although occasional "false positives" may be identified
by electrophoresis (9) of the very low density lipoprotein (VLDL)
fraction, therapy for a bonafide type V would not be significantly
different. Since sound medical practice involves screening members
of a type III's family with electrophoretic definition of type III,
some families may be screened unnecessarily.

It is agreed that it is possible to treat hyperlipoproteinemia
empirically on a hit and miss basis usually with success knowing
only the TG, TC and chylomicron levels. However, typing in many

cases will allow a more systematic initial therapy (10) with an anticipated earlier response as well as more accurate prognosis for
the patient.

It also should be conceded that the typing system has failed
to put all lipoprotein abnormalities neatly into six discrete genetic types. Moreover, under dietary and/or drug manipulation an
individual patient may exhibit more than one "type" at different
times. Also, there are degrees of severity of each type as well as
combinations of abnormalities, and these observations cannot be
adequately expressed by qualitative electrophoresis and plasma TG
and TC levels. What is fundamentally needed is the quantitative
levels of the major lipoprotein classes to provide an appropriate
measure of the "degree of abnormality"; for example, the degree of
II'ness or of IV'ness. In addition, such a system could readily
handle and quantitatively characterize "mixed hyperlipoproteinemia"
expressing, for example, both II'ness and IV'ness. Further, these
abnormalities need to quantitatively relate to established normal
population values for each lipoprotein class (11). For example,
the percentile rank and standard deviation score value of each
lipoprotein class can appropriately identify the degree of lipoprotein abnormality as well as the degree of II'ness, IV'ness, etc.
Thus, quantitative internally standardized lipoprotein electrophoresis providing this information can add a needed new dimension to
the typing system. However, its potential usefulness can only be
determined after a reasonable period of application and usage.

ADVANTAGES OF AN AUTOMATED MICROELECTROPHORESIS SYSTEM

Tedious manipulations and calculations have been a major ob-
stacle in the clinical application of quantitative lipoprotein electrophoresis. To overcome these difficulties an automatic system
for the microdensitometry and calculations as part of a quantitative agarose gel electrophoresis system has been developed (12).
Quantitative results are internally standardized by serum cholesterol measurements using the mean wt % content of cholesterol in
each major lipoprotein class (5). The hardware, described in detail elsewhere (12), consists of microdensitometer, an analog to
digital converter, a cathode ray tube terminal, a teleprinter and
a PDP 8/e computer equipped with a disc. The complete system is
shown schematically in Fig. 1. Scans are made of electrophoretic
patterns of serum alone or in combination with the 1.006 g/ml VLDL
top and/or bottom preparative ultracentrifuge lipoprotein fractions
(13). The analysis corrects for baseline drifts, pre-beta asym-
metry and will properly identify and quantify the amount of VLDL,
LDL and HDL with corrections for "sinking pre-beta" and "floating
beta" in LDL and VLDL, respectively. The latter corrections require the use of the ultracentrifuge fractions. Final results are
given in mg/100 ml as well as percentile and standard deviation

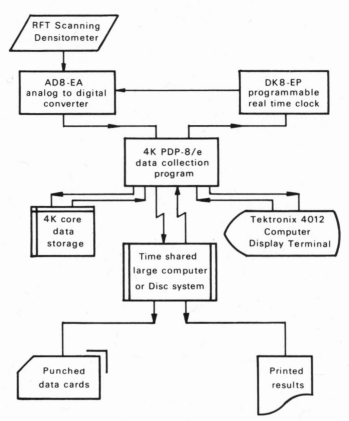

Figure 1:   Schematic diagram showing the functional units and the
flow of data in the automated system for quantitative lipoprotein
electrophoresis.   Reprinted from reference 12 by permission of the
Journal of Laboratory and Clinical Medicine.

score rank of each lipoprotein class as compared to an appropriate
normal reference population (11).   Examples of such data with ranking for two subjects with similar lipid levels are given in Fig. 2.
Such data are in a form more meaningful to the physician and patient
and may provide a quantitative dimension to lipoprotein phenotyping.

The above procedure inclusive of all calculations and appropriate ranking requires about 5 minutes technician time.   Further,
it is anticipated that in routine screening only a small fraction,
say 10%, of the samples will require the complete procedure involving lipoprotein fraction scans.   The plasma scan alone will be sufficient in most cases for lipoprotein quantification and ranking.
Given such conditions, two technicians could process some 250 serum
samples per week of which perhaps 25-36 might involve preparative

CASE 7184   SLIDE   350   AGE   48   SEX F
                                  G 900
CLASS	SU	MGPCT	ANUC	VS	27-45 FEMALES		46-65 FEMALES		ALL	FEMALES
					RANK	S.D.	RANK	S.D.	RANK	S.D.
HDL	14983	574.4	(537.5)		99%	2.31	97%	1.85	98%	2.04
VLDL	5255	94.4	(74.9)		84%	1.01	34%	-.40	45%	-.12
LDL	24692	540.9	(481.8)		97%	1.94	80%	.86	89%	1.24
TG		108.0			95%	1.64	38%	-.30	51%	.02
CS		289.0			100%	4.07	91%	1.34	97%	1.93

FRACTIONS UTILIZED   WS ONLY   STANDARDIZED USING   CS ONLY
ALPHA STAINING FACTOR= 1.75

CASE 7188   SLIDE   345   AGE   45   SEX F
                                  G 900
CLASS	SU	MGPCT	ANUC	VS	27-45 FEMALES		46-65 FEMALES		ALL	FEMALES
					RANK	S.D.	RANK	S.D.	RANK	S.D.
HDL	3522	178.5	(227.1)		10%	-1.30	5%	-1.61	8%	-1.43
VLDL	3946	93.7	(74.3)		84%	.99	34%	-.41	45%	-.12
LDL	25919	750.5	(666.7)		100%	4.47	100%	2.72	100%	3.26
TG		111.0			96%	1.79	39%	-.27	52%	.06
CS		292.0			100%	4.20	92%	1.41	98%	2.00

FRACTIONS UTILIZED   WS ONLY   STANDARDIZED USING   CS ONLY
ALPHA STAINING FACTOR= 1.75

Figure 2: Teleprinter output format with ranking for cases 7184 and 7188. Note that with similar TC and TG levels the HDL and LDL levels are substantially different. Abbreviations are: SU, scanning units; WS, serum; S.D., standard score of values with respect to our reference population (11) expressed in standard deviation units; ANUC, analytic ultracentrifugal values calculated from regression formulae (11) using TC values alone. Thus, ANUC values for HDL = 87 + 0.784x, VLDL = -6 + 0.859x and LDL = 5 + 0.882x, where x values are the calculated agarose concentrations internally standardized by TC alone.

ultracentrifugation. Without the centrifugation some 400-500 serum samples per week could be analyzed. Considering the above, a rough estimate of the unit cost exclusive of the serum cholesterol and triglyceride determinations would be approximately $5.00 - $10.00 per test.

Although we use a specific configuration of hardware, others should work as well or better. Also, with slight modifications, different systems should be able to utilize the substantial software developed for our system. Without going into detailed logistics of large scale screening, the average sample will require 3-4 minutes of processing which is a throughput potential comparable to a typical clinical autoanalyzer.

OTHER SIMPLIFIED QUANTITATIVE LIPOPROTEIN MEASUREMENTS

Alternatives to quantitative electrophoresis include two quantitative lipoprotein procedures based upon lipoprotein precipitation by heparin-$MnCl_2$. The procedure utilized by the Lipid Research Clinics (14) first removes VLDL by preparative ultracentrifugation, then precipitates LDL and calculates all lipoprotein concentrations in terms of cholesterol content of VLDL, LDL and HDL. This procedure does not lend itself to large scale application. A more simplified procedure by Friedewald et al. (15) eliminates the centrifugation step and estimates VLDL cholesterol directly as TG/5. However, samples with TG > 300 cannot be properly analyzed. After both LDL and VLDL are precipitated, LDL and HDL cholesterol are determined. Thus, the Friedewald technique also has the capability for large scale quantitative lipoprotein screening. However, limitations of both precipitation procedures are the necessity of performing accurate cholesterol analyses on dilute lipoprotein containing solutions.

Recently the Friedewald procedure and our automated microelectrophoresis procedure were compared (16) using the analytic ultracentrifuge as a standard (17). Thirty seven normal adult women were studied. Correlation coefficients relating ultracentrifugal and the Friedewald lipoprotein data were 0.849, 0.914 and 0.895 for HDL, VLDL, and LDL, respectively. Analogous correlations for quantitative electrophoresis were 0.929, 0.890 and 0.892, respectively. Table I gives a more detailed comparison of the two methods, including regression formulae for both procedures allowing calculation of equivalent ultracentrifuge data. Comparison of the error in estimating the assumed true value (the analytic ultracentrifuge results) is given by Sy.x. Although agarose provides a somewhat better estimation for HDL, no definitive advantage in accuracy is provided by either method.

However, there are advantages and limitations to both procedures. The Friedewald procedure can be used on frozen sera but requires 3 ml of sera. In contrast, the electrophoresis system must be used with fresh serum but requires only some 100 µl serum if an enzymatic autoanalyzer technique is used for the cholesterol and triglyceride determinations. Thus, such an electrophoresis system would be most useful in screening young children or in studies where only small blood samples can be obtained. Another consideration is that the electrophoretic slides may be quantitatively analyzed several years retrospectively. In balance, it would appear that the electrophoretic method offers advantages for a mass screening program and is comparable in accuracy to an alternate chemical precipitation procedure.

Table I

Comparison of Electrophoretic and $MnCl_2$-Heparin

Procedures with Analytic Ultracentrifugation[a]

Variable	Mean ± SD mg/100 ml	r	$S_{y.x}$	b	a
Agarose HDL	438 ± 111	0.929	33	43	0.717
Agarose VLDL	77 ± 57	0.870	26	2	0.770
Agarose LDL	466 ± 138	0.892	51	27	0.702
$MnCl_2$-Hep. HDL[b]	355 ± 93	0.849	47	78	0.785
$MnCl_2$-Hep. VLDL[c]	158 ± 67	0.914	21	-51	0.686
$MnCl_2$-Hep. LDL[d]	398 ± 106	0.895	50	-21	0.920

[a] y = b + ax, where y = analytic ultracentrifuge lipoprotein values and x values are agarose concentrations calculated by both TC and TG internal standardization (5).

[b] $HDL = 5.95 \times HDL_{TC}$.

[c] $VLDL$ = serum TG/5 x 7.30.

[d] LDL = (serum $TC-HDL_{TC}$ - TG/5) 3.01.

Reprinted from reference 16 by permission of the Americal Oil Chemists' Society.

THE FUTURE FOR QUANTITATIVE LIPOPROTEIN ELECTROPHORESIS

Certainly now is not the time to abandon electrophoresis. It is rather a time to increase efforts to improve lipoprotein quantification by this useful simple technique, and to test its potential for improving the definition and treatment of lipoprotein abnormalities. Such improved electrophoretic procedures should enhance diagnostic capability.

The case for intensifying rather than minimizing interest in lipoprotein electrophoresis can be summarized as follows:

If in only 10% of screening cases additional information is provided by quantitative electrophoresis, it is still worthwhile to consider electrophoresis on all cases, because a priori one cannot identify all those 10% of cases that would benefit from electrophoresis. An analogous medical situation is that of remotely suspected bone fractures. A physician would prudently recommend an x-ray in order to properly identify those very few patients who might have bona fide fractures.

Quantitative lipoprotein electrophoresis can provide concentrations of the major lipoprotein classes and these procedures and calculations can be conveniently automated (12).

Complete lipoprotein information is more meaningful and important metabolically than plasma triglyceride and cholesterol information. It seems obvious that these lipoprotein structures that exist naturally in the bloodstream are more important as objects of study than the two major lipid classes that have been extracted from all the lipoproteins (atherogenic and non-atherogenic) and lumped together as a total entity. Lipoproteins are the actively metabolizing units and metabolic defects are reflected in the entire plasma lipoprotein pattern. It is, therefore, logical that the lipoproteins are the important biological units to be studied and quantified.

Last and not the least important is that widespread discontinuance of electrophoresis will have an inhibitory effect of "basic research" which presently is attempting to improve and quantify lipoproteins by electrophoresis. An additional unfavorable effect will be within the practical clinical realm which has been aided by "off the shelf" commercial electrophoresis kits. Since most businesses operate on relatively short-term market expectations, some or most of these kits may be permanently withdrawn from the market if the immediate forecast for electrophoresis is negative.

The blood lipoprotein system is a complicated one and this is emphasized by the present inability to fit all lipoprotein abnormalities neatly into six types. It would appear essential that an improved classification will necessarily involve quantification of all lipoprotein classes. The most inexpensive, convenient, and simple method to do this is by quantitative lipoprotein electrophoresis.

ACKNOWLEDGEMENTS

This work was supported in part by National Institutes of Health grant HL-18574-02 and the U.S. Energy Research and Development Administration. We thank Suzanne Pan for able computer analysis of data, Gerald L. Adamson for excellent technical work and Dorothy S. Sprague for help in preparation of the manuscript.

REFERENCES

1. Beaumont, J. L., Carlson, L. A., Cooper, G. R., Fejbar, Z., Fredrickson, D. S. and Strassen, T. Bull. Wld. Hlth. Org. 43: 891, 1970.

2. Immarino, R. M. Human Path. 5: 626, 1974.

3. Fredrickson, D. S. Circulation 51: 209, 1975.

4. Dyerberg, J. and Hjorne, H. Acta Med. Scand. 191: 413, 1972.

5. Hatch, F. T., Lindgren, F. T., Adamson, G. L., Jensen, L. C., Wong, A. W. and Levy, R. I. J. Lab. Clin. Med. 81: 946, 1973.

6. Miller, G. J. and Miller, N. E. Lancet 1: 16, 1975.

7. Castelli, W. P., Doyle, J. T., Gordon, T., Hames, C., Hulley, S. B., Kagan, A., McGee, D., Vicic, W. J. and Zukel, W. J. Circulation 52 Suppl. II: II-378, 1975.

8. Wood, P. D., Stern, M. P., Silvers, A., Reaven, G. M. and von der Groeben, J. Circulation 45: 114, 1972.

9. Fredrickson, D. S., Morganroth, J. and Levy, R. I. Ann. Intern. Med. 14: 815, 1975.

10. Fredrickson, D. S., Levy, R. I., Bonnell, M. and Ernst, N. DHEW Publication No (NIH) 73-110, 1973.

11. Lindgren, F. T., Adamson, G. L., Jensen, L. C. and Wood, P. D. Lipids 10: 750, 1975.

12. Wong, R. A., Banchero, P. G., Jensen, L. C., Pan, S. S., Adamson, G. L. and Lindgren, F. T. J. Lab. Clin. Med., in press.

13. Lindgren, F. T.   In: <u>Analysis of Lipids and Lipoproteins</u>,
     E. G. Perkins, ed., J. Amer. Oil Chem. Soc., Champaign, IL.,
     1972, pp. 187-274.

14. Manual of Laboratory Operations, Lipid Research Clinics Pro-
     gram, Vol. 1, Lipid and Lipoprotein Analysis DHEW Publication
     No (NIH) 75-628, 1975.

15. Friedewald, W. T., Levy, R. I. and Fredrickson, D. S.   Clin.
     Chem. 18: 499, 1972.

16. Lindgren, F. T., Silvers, A., Jutagir, R., Layshot, L. and
     Bradley, D. D.   Lipids 12: 278, 1977.

17. Lindgren, F. T., Jensen, L. C. and Hatch, F. T.   In: <u>Blood
     Lipids and Lipoproteins</u>, G. J. Nelson, ed., John Wiley-Inter-
     science, N.Y., 1972, pp. 181-274.

SOLUTION PROPERTIES OF APOLIPOPROTEINS

Gerhard M. Kostner

Institute of Medical Biochemistry
University of Graz
A-8010 Graz, Harrachgasse 21, Austria

INTRODUCTION

We have learned from Dr. Lindgrens presentation that human serum
lipoproteins are prepared customarily by preparative ultracentrifugation. Quantitation of individual density classes can than be performed with the analytical instrument. Thus we are so far only
familiar with the operational classification of lipoproteins based
on the method used for their separation, reflecting the hydrated
densities. As it became evident from numerous publications of the
past, and this has been clearly demonstrated at the Protides Meeting
of this year, there exist a great variety of additional isolation
and subfractionation procedures for serum lipoproteins. In fact we
are able now to subfractionate lipoproteins not only according to
their flotation behaviour in salt solutions of increased densities,
but also according to differences in the protein moiety. Before
going into the sometimes troublesome matter of purifying apolipoproteins it proved to be of great advantage to separate first lipoprotein subspecies or families. For that purpose it is necessary to
apply not only one or two consecutive separation steps but rather a
combination of some 3-6 different ones. After isolation of a maximum
number of subfractions differing from each other by one or more apolipoprotein polypeptides they should be delipidated and further
processed. This is particularly true since apolipoproteins once
present in the lipid free state, tend to aggregate not only each
apoprotein with the same species but also with all kinds of polypeptides from other families. Thus aggregates are formed with high
molecular weights which in many cases cannot be dissociated
quantitatively. Part of this presentation will therefore be devoted
to the subfractionation of individual lipoprotein classes on a pre-

parative scale followed by the description of the most effective
and convenient delipidation procedures. Only the last part of this
chapter deals with the solubility properties of individual apolipoprotein polypeptides - a matter which today is only poorly investigated. Before going into this field let us just summarize the
present state of knowledge concerning the identity of apolipoprotein polypeptides of human serum.

TABLE I

Apolipoproteins of Human serum

Apo-lipoprotein	Synonym	Density Class	Mol.wt. x 10^{-3}	References
AI	R-GlnI	HDL	28	(1,2)
AII	R-GlnII	HDL	17	(1,2)
AIII	D	HDL	21	(3,4)
B	Apo-ß	LDL,VLDL	275	(5)
CI	R-Ser	VLDL,HDL	7	(6)
CII	R-Glu	VLDL,HDL	8.5	(6)
$CIII_1$	$R-Ala_1$	VLDL,HDL	8.5	(6)
$CIII_2$	$R-Ala_2$	VLDL,HDL	8.5	(6)
D	AIII	HDL	21	(3,4)
E	A.R.P.	VLDL,HDL	39	(7)
a	apoLp(a)	HDL_1	250-300	(8,9)
GRP	---	HDL	?	(10)
D-2	---	HDL	7	(11)

The family classification system proposed by Alaupovic (12) is
nowadays widely accepted to denominate lipoproteins and apolipoproteins. Since not all polypeptides of human serum lipoproteins
described so far can be integrated into this scheme, trivial names
must be used in addition in this report. According to the family
concept, apolipoproteins belonging to one family are characterized
by their simultaneous occurance into one lipoprotein particle.
Triglyceride rich lipoproteins form aggregates of lipoprotein
families. Above a hydrated density of approximately 1.025 or after
removal of neutral lipids, lipoprotein entities can be prepared
which meet more or less closely the requirements demanded for
"families". It should however, be emphasized that the in vivo
situation or the physiological significance of the described lipoprotein families have never been unequivocally demonstrated so far.

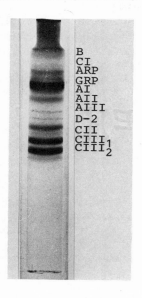

B
CI
ARP
GRP
AI
AII
AIII
D-2
CII
CIII$_1$
CIII$_2$

Fig. 1: Disc electrophoresis of 0,1 mg
of apoVLDL plus apoHDL in a
10% gel containing 8M urea.
Staining was performed with
Coomassie blue.

The two peptides listed at the bottom of Tab. 1 have only been
partially characterized and will therefore not be considered in
this report.

The best characterization of known or even new polypeptides is
the determination of amino acid composition and of terminal amino
acids. Since this characterization cannot always be carried out,
apolipoproteins are most conveniently characterized by their migration
rate in defined polyacryl amide gel systems and/or by their immunochemical properties. Fig. 1 shows the polyacryl amide gel electrophoresis pattern of a mixture of apo-VLDL plus apo-HDL. The apoproteins were electrophoresed in 10 % polyacryl amide gels (PAGE)
containing 8M urea. The particular gel composition has been
described in earlier publications (2,3) and with some experiance
there are little difficulties to correlate the stained bands with
the individual apolipoproteins. In addition, monospecific antisera
which unfortunately are not commercially available proved to be of
great importance to characterize and also quantitate polypeptides
(13). This is particularly true for apoB, the major constituent of
LDL, which does not migrate into the separating gels using the system
described for Fig. 1. Since the immunochemical behaviour of intact
lipoproteins and their lipid free counterparts may vary considerably,
all immunochemical characterizations should be performed with both
materials.

Isolation and Subfractionation of Serum Lipoproteins

Although it is not the aim of this presentation to introduce
the preparation of lipoprotein subfractions, this matter will still

be overviewed in order to be able to follow the subsequent explana-
tions.   Intact lipoproteins should be fractionated as far as possible and the separation of apolipoproteins should proceed with the
obtained material. The first step for preparing large amounts of
apolipoproteins most commonly are precipitation methods. Although
a great variety of substances have been described for more or less
selective precipitation of lipoproteins, only few of them can safely
be applied without alterations of the obtained material. From all
the precipitating agents described by Burstein (14) I feel that
sodium phosphotungstate can be recommended mostly for precipitation
individually VLDL, LDL and HDL. After solubilisation of the precipitate in O.1M sodium citrate buffer, lipoproteins are further
purified by preparative ultracentrifugation. Another very promising
procedure, recently introduced by Garcia (15) fractionates individual
lipoprotein classes by ammonium sulfate precipitation followed by
sodium phosphotungstate. In this way the preparative ultracentrifuge
can be completely avoided. Both methods yield lipoprotein mixtures
which are essentially free of nonlipoprotein components. A further
subfractionation of these is possible by many selected procedures.
The most effective ones are immunoadsorbers, lectines linked to a
solid matrix e.g. Sepharose 4B, and preparative column chromatography using Biogels or hydroxyapatite. By these methods, we and
others were able to subfractionate intact lipoproteins into numerous
distinct entities differing from each others not only by the lipid
but also by the protein moiety. Some of the subclasses represent
lipoprotein families characterized by the presence if apolipoproteins belonging to only LpA, LpB or LpC. As an example I briefly
will describe the method which has been successfully applied in our
laboratory for the isolation of the three major lipoprotein families.

     Isolation of lipoproteinB (LpB). Although LpB represents more
than 95 % of the lipoprotein mass of LDL, it is sometimes very
cumbersome to prepare LpB uncontaminated by apoA and apoC polypeptides. Yet a highly purified product is absolutely necessary
for studying the structure of apoB. In our procedure we start with
pooled serum from fasting normals. Immediately after sampling,
preservatives and protease inhibitors (Na_2EDTA, NaN_3, Merthiolate,
Chloramphenicol and phenyl methyl sulfonyl fluoride) should be
added. In order to avoid troubles expressed by the formation of
aggregates it is of advantage to start with serum from Lp(a) negative donors. After centrifugation in a laboratory centrifuge VLDL
and LDL are precipitated out from the serum according to Burstein
and Morfin (16). The precipitate is removed at 8000 g, washed once
with saline containing sodium phosphotungstate and $MgCl_2$ of the
same ratio as used in the first step, and suspended in 1% sodium
citrate. This suspension is dialysed overnight against saline
preferentially at $4^{O}C$. All solutions used throughout the isolation
procedure should contain some of the preservatives listed above.
The density of the dialysed solution is adjusted to 1.025 by the
addition of solid NaCl and VLDL and LDL_1 are floated in the

preparative ultracentrifuge at 120 000 g for 18 hrs and 15°C. The
sedimenting material is removed by tube slicing and adjusted to a
density of 1.055 with NaCl.LDL are floated at 125 000 g for 22 hrs
and 15°C.   The yellow layers on top of each tube are mixed, diluted
with Tris.HCl buffered saline (pH 8.5) to a final LDL concentration
of 10 mg/ml and chromatographed in portions of 8-10 ml over a
column, 100 x 2.5 cm, packed with Biogel A-5m in Tris-HCl buffered
saline. The elution pattern of a representative experiment is
shown in Fig. 2. The main peak contains LpB with a purity
exceeding 98.5 %, while peak 1 are aggregates and peak 3 LpA and
LpC. The use of Tris.HCl buffer seems to be essential for removal
of most of the apoA and C peptides since other substances as for
example Barbital-Na or $NaHCO_3$ were unable in our hands to
dissociate the LpB - LpA - LpC - complex.

   Isolation of LpA and LpC from HDL. LpA is the major lipoprotein
family of HDL. This density class is therefore used for the LpA
preparation. In most of the cases we start with HDL_3 as this
material represent more than two third of the total HDL. In
addition HDL_2 does not always yield pure LpA. HDL is prepared from
the supernatant of the VLDL-LDL precipitation step by ultracentrifugation at d = 1.22, adjusted with solid NaBr, for 24 hrs

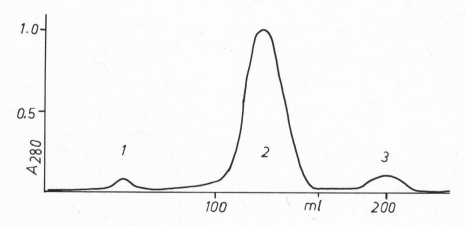

Fig.2: Elution pattern of low density lipoproteins
         (d 1.025 - 1.055) from a column, 2.5 x 100 cm packed
         with Biogel A-5m in 0.15M NaCl containing 0.05M/l
         Tris.HCl, pH 8.5. 100 mg LDL were applied.

at 145 000 g and 15°C. If larger volumes have to be processed, HDL
are precipitated also with sodium phosphotungstate (16). Either HDL
preparation is dialysed against a NaBr solution of d = 1.110 and
ultracentrifuged at 145 000 g for 30 hrs. The bottom fraction (two
third of the tubes) represent HDL$_3$ while HDL$_2$ floate under these
conditions. Since HDL$_3$ is the major substrate for the enzyme
lecithin:cholesterol acyl transferase (LCAT) we add to the starting
serum the inactivator Ellman's reagent (10mM/l). HDL$_3$ is dialysed
against 0.05 M potassium phosphate buffer (KPP), pH 6.8, and passed
over a column packed with hydroxyapatite in the same buffer.
Hydroxyapatite prepared according to the procedure described at the
Protides meeting (17) should be used in order to obtain good flow
rates. LpA, representing more than 60% of the applied material,
elutes with the 0.05 M KPP. With 0.1 - 0.2M KPP some aggregation
products of LpA with LpC are eluted while a lipoprotein with only
apoC peptides is obtained with 0.25 - 0.35 M KPP. The so obtained
lipoprotein families are used as starting material for preparation of apolipoproteins.

Preparation of Apolipoproteins

    Since the solubility properties of apolipoproteins are reflected
most drastically by the delipidation procedure applied, this matter
will be covered here in more detail. An enormous variety of
delipidation methods have been described in the literature and there
are probably only few of them which have not been tried in my
laboratory. It tourned out that only the minority of them can be
recommended for general use. I will restrict my presentation to
those only which I feel are the most convenient and effective ones.
A more detailed review of this subject can be found in Ref. (18).
In general, two different possibilities exist to obtain apolipoproteins: Delipidation of lipoproteins in solution and delipidation
of freeze dryed material.

    Delipidation of lipoproteins in solution. These methods yield
apolipoproteins with very good solubility properties and are
particularly recommended for preparation of apoB. A very safe method
is that described by Shore and Shore (19) using diethylether and
ethanol -diethylether at 0-4°C. Lipoproteins are dialysed against
0.5 M NaCl, 0.01 M Tris, pH 7.6, 0.0008 M EDTA and extracted first
with diethyl ether and than with ethanol: diethyl ether 2:1, until
a lipid free apolipoprotein suspension is obtained. The major
drawbacks of this method are the solubility of the small peptides
especially of the C- family in the organic phase and the
accumulation of large volumes if greater amounts of lipoproteins
are delipidated. Another possibility for delipidation is the use
of detergents. Helenius and Simons (20) compared the effectivity
of Nonident P40, Triton x 100, sodium dodecyl sulfate (SDS) and

Na-desoxycholate (NaDOC) for the preparation of apoB. The best
results were obtained with NaDOC. 11.5 mg of this substance are
added to each mg of LDL at pH 10, adjusted with Na_2CO_3, and the
mixture is chromatographed over Sephadex G200 or Biogel. The first
peak represents apo-LDL while lipid micelles are eluted in a later
fraction.

Another very efficient method is that described by Kane (21)
using tetramethyl urea (TMU), especially if small amounts of lipoprotein solutions are to be delipidated. Although this method
delipidates quantitatively only VLDL and LDL, it can be used also for
HDL if samples are investigated in PAGE in the presence of 8M urea.
The lipoprotein solutions are mixed with equal volumes of TMU and
immediately electrophoresced in 8M urea containing gels. ApoAI may
form aggregates if higher concentrations are present, which
eventually interfere with the peptide band pattern. This method
has also been successfully applied to quantitate the apoB content
of VLDL or LDL in metabolic investigations. ApoB namely precipitates
out quantitatively under the described conditions while ApoA and
apoC remain in solutions, thogether with the lipid part. Though
it is not possible to separate all peptides from the lipid moiety
with TMU.

To circumvent these problems, we searched for delipidating
agents which are non-miscible with aqueous buffers and separate
lipids from proteins in an organic phase. One of the most powerful
effective and milde substance with that respect was found to be
glycerol-1-hexyl ether (GHE). The synthesis of GHE has been
described in detail at the last year EFRAC meeting in Brugge   (22):
Isopropylidene glycerol and hexyl bromide both in benzene are
mixed at a molar ratio of 2:1 and boiled under reflux in a water
separator. Solid KOH is added after one hr and the whole mixture
is poored on crashed ice. The formed GHE is extracted with diethyl
ether and distilled under reduced pressure (BP at 0.2 mm Hg: $88^{\circ}C$).
GHE is a colorless, harmless liquid which can be stored at room
temperature for years without loss of its property. Upon the addition
to lipoprotein solutions, GHE removes immediately all lipids from
VLDL and LDL. HDL are only partially delipidated. Pretreatment of
HDL with phospholipase-C yields also totally delipidated apoHDL.

Delipidation with GHE. For delipidation of VLDL, LDL or LpB,
0.1 - 10 ml of solution containing 1-30 mg of lipoproteins per ml
of 0.15 M NaCl are mixed at room temperature with 10 % of the volume
with GHE and shaken for 30 min. vigorously with a Vortex mixer.
The mixture is centrifuged for 10 min. at 5000 g and the organic
phase at the top is aspirated by a syringe. Addition of diethyl
ether at this step may save 50 % of the GHE. The extraction procedure
is repeated once or twice. The formed apoB accumulates at the inter-
phase while apoA and apoC peptides remain mostly in solution. HDL
are preincubated at an enzyme : substrate ratio of 1:2000 for

12 hrs at $37^\circ C$ in the presence of 0.05 M Ca^{++} and delipidated as described for LpB. The described method handles any amount of lipoprotein and works extremely rapid. It is recommended primarily for delipidation of LpB and for serial analyses of any lipoprotein fraction in small volumes.

In addition to the methods described here for delipidation of lipoproteins in solutions, a great deal of other procedures have been described in the literature. Some authors work with isobutanol: acetic acid: water mixtures and chromatograph this mixture over Sephadex LH 20, others are using chloroform methanol. We however consider the GHE, NaDOC and TMU methods as the most effective ones.

Delipidation of freeze dryed lipoproteins. For some purpose the delipidation of lyophilized lipoprotein fractions should be prefered to the methods described in the former paragraph. The major advantages are that all apolipoproteins definitely remain insoluble in the organic phase. In addition the accumulation of large volumes can be avoided if one works with large amounts of material. It should, however be noted that apoB in these cases is very hard to solubilize even in buffers containing detergents. We have successfully applied the following method for delipidation of chylomicrons, VLDL and HDL: Lipoprotein fractions are dialysed eshaustively against a solution of 0.001 M $NaHCO_3$ containing 0.01 mg/ml of Na_2EDTA. The volume and concentration of the solution is not crucial. After equilibration with the buffer in the cold room the solution is lyophilized. Triglyceride rich lipoproteins but also huge amounts of HDL are extracted first with peroxide free diethyl ether at room temperature. This step eventually is performed in a Soxhlet apparatus. Alternatively centrifuge tubes with glass stoppers are used. The extraction with diethyl ether is repeated several times until an aliquot of the extract gives no residual lipid after evaporation. In the next step ethanol: diethyl ether 3:1 is used. This extraction step has to be carried out in the cold room and care has to be taken that by no means the temperature of the lipoprotein raises over $4^\circ C$. The amount of solvent and number of extraction steps must be adapted individually. HDL or LpA has to be extracted 5-6 times, VLDL and chylomicrons 3-4 times. After the removal of all polar lipids two to three more extractions with diethyl ether follow. The obtained white material is dried under nitrogen in a desiccator. Water free apolipoproteins can be stored at room temperature or $4^\circ C$.

Antigenicity of Delipidated Apolipoproteins

For the location of apolipoproteins during their separation but also for their determination in individual lipoprotein subfractions immunochemical tests are widely used in addition to other analytical methods as for example PAGE. There seems to be little

doubt that the immunochemical behaviour of most if not of all polypeptides differ after delipidation from that in intact lipoproteins.
This phenomenon has clearly been demonstrated for example with
apoAI, which is displaced in the radio-immunoassay by the corresponding amount of LpA only to 5 % (23). In addition it has been
shown that the C- and N-terminal region of that apolipoprotein has
different antigenic properties. For quantitative immunochemical
tests therefore, variable results can be obtained with antisera
from different species. For qualitative tests all apolipoproteins
are solubilized best in 0.05 M Tris.HCl, pH 8.0, containing 8 M
urea and investigated by immunoelectrophoresis or immunodiffusion.
The shape of the precipitin arc in immunoelectrophoresis using a
specific antiserum may vary from intact to delipidated lipoproteins.
In immuno diffusion experiments however for all apolipoproteins a
reaction of complete identity has been obtained in our laboratory.
ApoB solely, gives only immunochemical reactions in the presence
of certain detergents as for example SDS or NaDOC. With Triton X100
on the other hand-though capable to solubilize apoB - no precipitation using anti-LpB has been observed. Applying SDS another
unspecific precipitate in addition to the specific precipitine line
is formed which is likely to mislaed unexperienced researchers.

Solution Properties Of Apoproteins

        As pointed out above, the properties of apoproteins depend to
a major degree upon the isolation and delipidation procedures.
Generalizing matters on can say that all polypeptides investigated
so far get more and more insoluble if repeated freeze drying is
performed or if extreme pH values· are applied. In addition it has
been found that mixtures of nonidentical apolipoproteins in most
cases are much better soluble than purified polypeptides. The exact
condiiton for the most effective solubilization of all apolipoproteins individually is only poorly investigated. The statements
made in the following are therefore restricted to some limited investigations carried out in our laboratory or partly also published
by other research groups. The solubility properties of apolipoproteins can be drastically altered on chemical modification. The
most frequently used methods are acetylation, maleylation, succiny-
lation and citraconylation. The latter has the particular advantage,
that the covalently linked ligands can be removed quantitatively in
a short period of time at pH 4.0 or bellow. The following report,
however, deals only with untreated apoprotein samples. The solubility
properties in addition are influenced markedly by the presence of
residual lipids-especially by phospholipids. This seems to be the
main reason why divergent reports have been published from several
laboratories. In our studies we have checked the purity of apolipoproteins for residual neutral lipids and phospholipids. The phospholipid content in all cases was less than 0.4 % by weight while other
lipids were absent.

   Solubility of apoB. Apolipoprotein-B requires the strongest
detergents for solubilisation. This seems to be the reason why the
exact number and molecular weight of apoB polypeptide is still a
matter of speculation. What makes matters worse is the fact that
apoB is prone to enzymatic degradation by proteases of endogenous
or bacterial origins. The degraded material has other solubility
properties as the intact B-protein.

   ApoB is very sensitive to the delipidation methods. After
delipidation in solution applying the methods described above, apoB
gives clear solutions in organic acids such as 50 % acetic acid,
80 % formic acid or others. Centrifugation of these products at
500 000 g min., however, yields pellets of aggregated apoB. The
supernatant contain small amounts of possibly "monomeric" apoB. In
the presence of large amounts of A or C-peptides, apoB can even be
solubilised in 8M urea or 6M guanidinium hydrochloride (Gu.HCl).
Much better solubilisation of apoB is achieved with SDS, sodium
decyl sulfate (SDeS) or NaDOC. If the latter detergent is used the
delipidation method of Helenius and Simons has to be applied (20).
With SDS highly concentrated apoB solutions exceeding 20 mg/ml can
be obtained. The presence of reducing agents e.g. mercapto ethanol
or DTT is not obligatorybut helps to avoid aggregation. To 1 mg of
apoB 8 mg of SDS must be added in addition to the desired amount of
buffers. The mixture is heated for 3 min at 100°C for inactivation
of proteases and stirred under nitrogen until a clear solution is
formed. Dialysis can now been applied against buffers containing

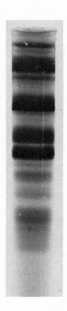

Fig. 3:

Polyacryl amide gel electrophoresis
of apolipoprotein-B in a 5 % gel
containing 0.1 % SDS. Staining was
performed with Coomassie blue.

0.1 % SDS or 0.01 % SDeS. Further removal of detergents by dialysis leads to the precipitation of apoB. Fig. 3 shows the PAGE pattern of such apoB in 5 % gels, 0.1 % SDS. Although all mentioned preservatives were present during preparation it cannot be excluded that we are dealing here with a secondary pattern created by proteases.

Solubility properties of apoA peptides. The major work concerning LpA has been performed on apoAI and AII and no systematic analyses on apoAIII are known. ApoAI and most probably also AIII are single stranded polypeptides while AII is composed of two identical subunits linked with disulfide bonds. ApoAI and AII have been found to undergo selfaggregation in all kinds of buffers and solvents. From their role lipid binding proteins one would expect that these interactions are caused by hydrophobic forces. This assumption is underlined by the findings that anionic and cationic detergents bind to apoAI and AII with the same number of sites (i.e. 3-4) and with the same order of magnitude (association constants 2×10^4 litres/mol) (24), and further that saturation of the hydrophobic regions counteract the aggregate formations.
ApoAI and AII exhibit marked secondary and tertiary structures in ordinary buffers, which are temperature dependent and easy to destroy not only by detergents but also at exteme pH values i.e. bellow 2 and above 12, (25), or substances such as Gu.HCl or urea. From all these observations it is easy to understand that the solubility properties of apoAI and AII are mostly reflected by their formed structure which is subject to change under various experimental conditions. This process is not reversible under all circumstances. In Tris-HCl buffer, pH 8.3, ionic strenght 0.05, apoAI forms dimers and apoAII tetramers. Addition of Gu.HCl or other dissociating agents to apoAI leads to formation of the monomeric peptide. ApoAII in addition requires the presence of mercapto ethanol or DTT. Summarising the results one can say that apoAI and AII are well soluble in slightly alkaline buffers though tend to form aggregates of dimers or oligomers. By the addition of detergents or dissociating agents such as SDS, SDeS, Gu-HCl or urea, monomers are formed.

Organic acids have also successfully applied for the solubilisation of LpA peptides. 1-2 M acetic acid is a very effective solvent and was used for isolation of individual LpA polypeptides. At concentrations of 10 mg/ml or less most of the A- and C-peptides seem to be present in monomeric forms and can be conveniently separated over Sephadex G75 - G200 or other molecular sieves (2,3). Immediately after separation the fractions can be lyophilized. It should be noted here that only analytical grade reagents should be used throughout all investigations of apolipoproteins.
For apoAIII (apoD) essentially the same statements can be made as for AI and AII though exact investigations on this protein are missing so far.

Solubility properties of apoC peptides. All apoC polypeptides have a much lower molecular weight comparing to apoA. Without doubt apoC peptides do have also hydrophobic regions in their primary structure which have clearly demonstrated by Gotto's group (26). Detergents have therefore frequently been used for their purification. The hydrophobicity of C peptide might also be reflected by their solubility in organic solvents e.g. ethanol - diethyl ether - water mixtures. Profound investigations on the solubility of apoC peptides are not known so far. This report therefore is restricted to own observations. For apoC we made the same observations as for apoA namely that with the progress of purification their solubility gets worse. This is especially so in buffers containing no dissociating agents. Nevertheless in most cases monomeric forms seem to exist in buffers containing 6-8M urea of pH 7-9. Repeated freeze drying and other processing renders apoCI, CII and CIII in a form which tends to give aggregates in ordinary buffers. ApoC is excellently soluble in 1-2M acetic acid. Extreme pH values and detergents certainly change the secondary and tertiary structure of apoC in the same manner as observed for apoA. After removal of these substances aggregate formations are more likely to occur as compared to untreated apoC.

Solubility properties of other apolipoproteins. One major constituent comprising approximately 15% of the VLDL protein is the "arginin rich" polypeptide (A.R.P.) also called apoE. Nevertheless this peptide has been characterized only very recently (7). This most probably was caused by the fact that A.R.P. is almost insoluble in the absence of dissociating agents or detergents. Monomeric A.R.P. solutions may only be formed in the presence of 6M Gu.HCl, SDS, or possibly also 8M urea. We have repeatedly used 1 M acetic acid for isolation of A.R.P. over Sephadex G100. The latter two chemicals however seem to be to weak to be used for solubilization of isolated lyophilized A.R.P.
Another apolipoprotein which should be considered here is the "a" peptide the characteristic antigen of Lp(a). It differs widely from all proteins mentioned so far since it is a glycoprotein with more than 30 % carbohydrates by weight. The major portion of carbohydrates is neuraminic acid. The exact molecular weight of the "a" peptide is not known but it seem to exceed 250 000 daltons (9). Because of the high neuraminic acid content, the a-protein does not aggregate in ordinary buffers and neither urea nor detergents are necessary for its solubilisation. After delipidation of Lp(a) the antigen can be extracted with O.1M Tris.HCl, pH 8.2. A certain portion however, depending on the isolation procedure, remains complexed with apoB. This part cannot be extracted even not with the strongest dissociating agents. The pattern of apo-a in PAGE in the presence of 8M urea is essentially the same as without urea. In SDS containing gels the same kind of pattern is observed as compared to the former gel system. Much work remains to be done in order understand the

kind of interactions of apo-a with LpB giving rise to the formation
of the Lp(a) lipoprotein.

SUMMARY AND CONCLUSION

Human serum lipoproteins are constituted of a minimum of then
nonidentical apolipoproteins. Alignement of these peptides with
increasing solubility in ordinary buffers should give the following
picture: ApoB≪A.R.P. < apoAIII < apoAI < apoAII = apoCII < apoCI <
apoCIII ≪ apo-a. Excepting apo-a the same figure is obtained if
peptides are listed with decreasing molecular weight. It has been
found that the solution properties of almost all apoproteins is
mostly reflected by the procedures applied for their purification
and even more by the delipidation method. For the polypeptides
apoB, A.R.P. and possibly also apoAIII delipidation of lipoprotein
solutions using TMU, GHE or detergents are recommended to yield
fractions with optimal solution properties. Other peptides may be
delipidated with diethyl ether and ethanol out of solutions or of
freeze dryed lipoproteins. The quality of the delipidated material
in many cases can be easily checked by inspection of the pattern
in PAGE or the precipitin arcs in immunoelectrophoresis. The
occurance of single sharp bands without major staining of gel
portions between bands in PAGE suggest very good solubility. The
observation of smeary bands on the other hand indicates partial
alterations of the material. Generalizing matters on may say that
apolipoproteins obtained directly after mild delipidation are very
well soluble in most cases even in the absence of dissiciating
agents (excepting apo-B). As the purification of distinct polypeptides proceeds the obtained material tends to aggregate even in
the presence of urea or acetic acid. In ordinary buffers all the
apolipoproteins listed above excepting apo-a tend to aggregate,
forming dimers and oligomers. The forces causing this associations
are of hydrophobic nature but also hydrogen bonds are likely to
occur. Ionic detergents especially SDS or SDeS are capable to
render the aggregates in monomeric forms. Since the latter are not
easy to remove quantitatively, other substances are frequently
used. The strongest one in that sense seems to be Gu.HCl which gives
in all cases monomeric solutions excluding apoB. Less powerful but
still very frequently applied are 8M urea and 1-2M acetic acid. The
presence of residual lipid material, especially of lecithin, markedly
influences the behaviour of apolipoproteins.

The immunochemical properties of all apolipoproteins present
onto intact lipoproteins differs from that in the lipid free state.
ApoAI for example has twenty times more antigenic sites than LpAI.
Similar observations have been made with other peptides. Nevertheless all apoproteins with the exception of apoB do form immunochemical reactions of identity before and after delipidation. ApoB
has to be solubilised in NaDOC or SDS in order to obtain an immuno-

chemical reaction. Nonionic detergents, though capable to solubilise the B protein, are useless for that purposes. Although a great deal of progress has been reached in recent years much more work seems to be necessary to uncover the exact interactions of apolipoproteins with each other and also with different classes of lipids. Some of the methodology outlined in this report may help to do so.

Part of these investigations have been supported by the Österreichischen Forschungsrat and the Österreichischen Nationalbank.

REFERENCES

1) B.Shore and V.Shore, Heterogeneity in protein subunits of human serum high density lipoproteins, Biochemistry 7, 2773-2777 (1968)

2) G.Kostner and P. Alaupovic, Studies of the composition of plasma lipoproteins. C-and N-terminal amino acids of the two non-identical polypeptides of human plasma apolipoprotein A, FEBS Lett. 15, 320-324 (1971)

3) G.Kostner, Studies of the composition and structure of human serum lipoproteins. Isolation and partial characterization of apolipoproteinAIII, Biochim.Biophys.Acta, 336, 383-395 (1974)

4) W.J.McConnathy and P.Alaupovic, Studies on the isolation and partial characterization of apolipoprotein D and lipoprotein D of human plasma, Biochemistry 15, 515-520 (1976)

5) R.Smith, J.R. Dawson and C.Tanford, The size and number of polypeptide chains in human serum low density lipoproteins, J.Biol. Chem. 247, 3376-3381 (1972)

6) W.V.Brown, R.I.Levy and D.S.Fredrickson, Further characterization of apolipoproteins from the human plasma very low density lipoproteins, J.Biol.Chem. 245, 6588-6594 (1970)

7) B.Shore and V.G.Shore, An apolipoprotein preferentially enriched in cholesteryl ester rich very low density lipoproteins, Biochem. Biophys.Res.Commun. 58, 1-7 (1974)

8) G.Utermann and H.Wiegandt, Darstellung und Characterisierung eines Lipoproteins mit Antigenwirksamkeit im Lp-System. Humangenetik 8, 39-46 (1968)

9) G.Jürgens and G.M.Kostner, Studies on the structure of the Lp(a) lipoprotein. Isolation and partial characterization of the Lp(a) specific antigen, Immunogenetics 1, 560-574 (1975)

10) S.O.Olofsson and A.Gustafson, Studies on human serum HDL:II.Isolation of subfractions in the cold, Scand.J.clin.Lab.Invest. 34, 257-265 (1974)

11) C.T.Lim, J.Chung, H.J.Kayden and A.M.Scanu, Apoproteins of

human serum high density lipoproteins. Isolation and characterization of the peptides of Sephadex fraction V from normal subjects and patients with abetalipoproteinemia, Biochim.Biophys Acta 420, 332-341 (1976)

12) P.Alaupovic, Apolipoproteins and lipoproteins, Atherosclerosis 13, 141-146 (1971)

13) G.M.Kostner, J.R.Patsch, S.Sailer, H.Braunsteiner and A.Holasek, Polypeptide distribution of the main lipoprotein density classes separated from human plasma by rate zonal ultracentrifugation, Eur.J.Biochem. 45, 611-621 (1974)

14) M.Burstein, H.R.Scholnik and R.Morfin, Rapid method for the isolation of lipoproteins from human serum by precipitation with polyanions, J.Lipid Res. 11, 583-595 (1970)

15) M.Garcia, Darstellung des gesamten Spektrums der Humanplasma-lipoproteine durch Präzipitation, Thesis, University of Marburg/Lahn (1976)

16) M.Burstein and R.Morfin, Precipitation des Alpha Lipoproteins du serum par le Phosphotungstate de Sodium en presence du Chlorure de Magnesium, Life Sci. 8, Part II, 345-348 (1969)

17) G.M.Kostner, The distribution of apolipoproteins in subfractions of normal human serum lipoproteins obtained by various techniques, Prot.Biol.Fluids, XXV Coll. (1977)

18) J.H.Ledford, Thesis: Studies on the delipidation of human serum lipoproteins, Univ.of Oklahoma (1969)

19) V.G.Shore and B.Shore, Heterogeneity of human plasma very low density lipoproteins. Separations of species differing in protein components. Biochemistry 12, 502-507 (1973)

20) A.Helenius and S.W.Simons, Removal of lipids from human plasma low density lipoproteins by detergents, Biochemistry 10, 2542-2547 (1971)

21) J.P.Kane, A rapid electrophoretic technique for identification of subunit species of apoproteins in serum lipoproteins, Analyt.Biochem. 53, 350-364 (1973)

22) G.M.Kostner, Studies on the structure of lipoprotein-B: Preparation of a soluble apolipoprotein-B. In: Lipid-Protein Complexes,Proc. of the EFRAC Conference in Bruges 1966, Plenum press.

23) G.Schonfeldt, R.A.Bradshaw and J.S.Chen, Structure of high density lipoprotein. The immunologic reactivities of the COOH- and NH_2 terminal regions of apolipoproteins AI. J.Biol.Chem. 251,3921-3926 (1976)

24) W.L.Stone and J.A.Reynolds, The Self-association of the apo-GlnI and apo-Gln-II polypeptide of human high density serum

lipoproteins, <u>J.Biol.Chem.</u> 250, 8045-8048 (1975)

25) J.Gwynne, B.Rewer and H.Edelhoch, The molecular properties of
apoAI from human high density lipoprotein, <u>J.Biol.Chem.</u> 249,
2411-2416 (1974)

26) J.P.Segrest, R.L.Jackson, J.D.Morisett and A.M.Gotto,Jr. A
molecular theory of lipid-protein interactions in the plasma
lipoproteins, <u>FEBS Letters</u> 38, 247-253 (1974)

APOLIPOPROTEIN QUANTITATION

Peter N. Herbert,* Linda L. Bausserman,
Lloyd O. Henderson,* Robert J. Heinen,
Marguerite J. LaPiana, Eve C. Church, and
Richard S. Shulman

Molecular Disease Branch, National Heart, Lung, and
Blood Institute, National Institutes of Health,
Bethesda, Maryland
*Division of Clinical and Experimental Atherosclerosis,
The Miriam Hospital, Providence, Rhode Island

The problems encountered in attempts to quantitate the apolipoproteins are common to all efforts to quantify minor constituents of
complex mixtures. Considerations of selective and non-selective
losses in the preparation of test samples, specificity, sensitivity,
precision, and accuracy are applicable to the selection and evaluation of available techniques. The natural association of apolipoproteins with lipids presents additional problems that must be resolved in the validation of a proposed procedure.

Two general approaches to apolipoprotein quantitation will be
discussed in this chapter. Estimations based on column chromatographic fractionation and polyacrylamide gel electrophoresis will be
considered first. These methods require isolation of the lipoproteins before apoprotein quantitation. Immunological methods including single diffusion, electroimmunoassay and radioimmunoassay
will be emphasized because of their applicability to quantitation
in both whole plasma and lipoprotein subfractions.

TECHNIQUES APPLICABLE TO ISOLATED LIPOPROTEINS

Preparative Ultracentrifugation. Ideal techniques for lipoprotein isolation are not available. Isopycnic ultracentrifugation

in fixed angle rotors is most widely employed (1). Very low density
lipoproteins (VLDL) can be recovered by a single centrifugation step
at plasma density, but isolation of low density lipoproteins (LDL)
and high density lipoproteins (HDL) require a minimum of two ultracentrifugation steps at high salt concentrations. All lipoprotein
classes so isolated are usually contaminated with other plasma proteins that may be removed by one or more recentrifugations. Such
ultracentrifugal "washing," however, is attended by selective losses
of both lipids and proteins and alteration of lipoprotein physical
properties (2). Such washing is most useful when apoprotein purification is the goal. Removal of non-apolipoprotein contaminants
enhances the likelihood of obtaining individual apoproteins in homogeneous forms. When simple apoprotein quantitation is the goal,
however, ultracentrifugal washing is best avoided.

     Delipidation of Lipoproteins. The methods for apoprotein
quantitation discussed in this section require delipidation prior to
application of the quantitative test. This is usually achieved by
lipid extraction with organic solvents (3, 4), detergents (5), or
tetramethylurea (TMU) (6). When column chromatography has been employed for apoprotein semiquantitation, a variety of methods of organic solvent extraction have been advocated. The most widely used
is that recommended by Scanu and Edelstein in which a solvent system
of ethanol and diethylether is employed (3). In the authors' experience, the most reliable combination of organic solvents is that
of methanol and diethylether. Lipoproteins are dialyzed against
0.15M NaCl, 1 mM EDTA, 5 mM NH_4HCO_3. Two ml or less of the lipoprotein solution are added dropwise to 15 ml of vortexing methanol
at $0^{o}C$ in a 45-50 ml conical tube. The tube is filled immediately
with diethylether, inverted several times and placed for 10 minutes
in wet ice. The protein is sedimented by low speed centrifugation
(e.g., 2,000 rpm in any clinical centrifuge with swinging bucket
rotor) for two minutes. The organic solvent is removed by aspira-
tion or decantation, the protein resuspended in methanol, and the
tube again filled with diethylether and placed for 10 minutes in wet
ice. The protein is pelleted by centrifugation (this time 3,000 rpm
for four minutes) and the organic solvent again discarded. Finally,
the protein is rinsed twice with diethylether and dried in a thin
film on the tube surface with a stream of nitrogen. Residual traces
of ether are removed by vacuum suction before resolubilization.

     The most frequent reasons for inadequate delipidation, non-
quantitative recovery of protein, or recovery of protein with poor
solubility characteristics include: (1) failure to dialyze excess
KBr from the lipoprotein solution; (2) dialysis of the lipoprotein
solution against distilled water rather than 0.15M NaCl; (3) failure
to keep organic solvents at $0^{o}C$ or less; (4) use of prolonged or
high speed centrifugation to initially sediment the protein when the
delipidation mixture contains the original volume of water (the

conditions of centrifugation are much less important after the
initial protein sedimentation); and (5) drying the protein in a
pellet rather than a thin film. Methanol is the critical component
of this organic solvent mixture. Substitution of chloroform for
diethylether as the apolar solvent does not improve the degree of
delipidation (usually less than 1% residual phospholipid) and renders the protein much less soluble.

Apoprotein Solubilization. Techniques of apoprotein solubilization are discussed in detail elsewhere in this monograph and will
be treated only briefly here. ApoVLDL and apoLDL, when totally delipidated and in dry form, require detergents for total resolubilization. Sodium decylsulfate is optimal at a concentration of 0.1-
0.2M because it is more easily removed by dialysis. Sodium dodecylsulfate is not as readily dialyzed, but is satisfactory. (Both will
be referred to here as SDS). Detergent solubilized apoVLDL and apo-
LDL can be immediately fractionated by column chromatography in
buffers containing urea or low concentrations of SDS (e.g., 2 mM).
If chromatography in guanidine HCL-containing buffers is desired,
the anionic detergents must first be removed by dialysis because
high molarity guanidine HCl will precipitate SDS and some protein.

ApoHDL is readily solubilized in aqueous solutions containing
6-8M urea (7), 4-6M guanidine HCl, 50% acetic acid (8) or 0.1M
NH_4OH.   Solubilization in simple Tris buffers at pH 8.0 has also
been frequently used, but resolubilization in such buffers in sometimes slow and incomplete.

Purified apolipoproteins invariably are more difficult to solubilize than the same apoproteins present in mixtures. Storage in
solutions containing urea should be avoided because of the inevitable
carbamylation (9, 10), and guanidine HCl buffers are of too high ionic
strength to permit gel electrophoresis. We have found the following
solutions suitable for completely resolubilizing purified apolipoproteins in concentrated solution: 0.1M NH_4OH for apoproteins A-I,
C-II and E; 0.1M NH_4HCO_3 for A-II, C-III-1 and C-III-2; and 1.0M
acetic acid for C-I. When apoB from VLDL or LDL is required in
concentrated solution, the dilute protein solutions containing 2 mM
SDS, 0.1M NH_4HCO_3 are lyophilized, resolubilized in a small volume
of 0.1M NH_4OH and excess SDS removed by dialysis. The routine use
of volatile buffers permits repeated lyophilization for storage
without great increases in salt concentration.

Quantitation by Column Chromatography

ApoVLDL. Gel chromatography has been employed in many laboratories to obtain estimates of the quantities of major protein
species in VLDL. No chromatographic medium or buffer system permits

resolution of all of the VLDL apoproteins and the choice of chromatographic conditions is dictated in part by the fractions whose
quantitation is most relevant to the experimental protocol. ApoVLDL
can be resolved into two major and one minor peak on columns of
Sephadex G-150 (11) or G-200 (10) (Fig. 1) in buffers containing SDS.
The void volume peak (fraction 1, Fig. 1) contains mostly apoB, and
the C-proteins comprise the second major peak (fraction 3, Fig. 1).
The B protein is relatively pure, but the C-protein fraction is
typically contaminated with proteins from the minor peak (fraction
2, Fig. 1) which contains apoE and a number of other apoproteins and
contaminating plasma proteins (10).

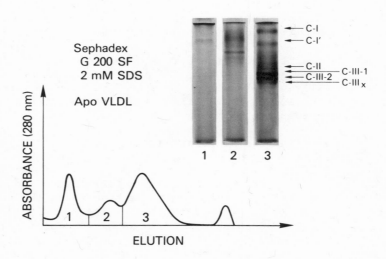

Fig. 1. Sephadex G-200 S.F. chromatography of 20 mg of apoVLDL. A
2.5 x 150 cm column was packed with the glass bead technique and
eluted in descending fashion (10). Polyacrylamide gel electrophoresis of the indicated pools was performed in 7.5% acrylamide monomer
in the system of Reisfeld and Small (12) modified so that all solutions contained 8M urea. S.M. refers to the starting material and
Arabic numbers to the corresponding chromatographic fractions.

Chromatography of apoVLDL on columns of Bio-Gel A 0.5M in 0.1M NH_4HCO_3, 2 mM SDS, pH 7.9, permits even better resolution of the apoB peak (fraction 1, Fig. 2). On this medium apoE is recovered in fractions 2 and 3, almost all apoC-I and considerable apoC-III-1 and apoC-III-2 in fraction 4, and all of the apoC-II and some apoC-III in fraction 5 (Fig. 2). This system is particularly useful for isolating apoC-II in preparation for ion-exchange chromatography, but the resolution of apoE is suboptimal.

ApoVLDL can additionally be resolved in a variety of chromatographic media eluted with guanidine-HCl or urea-containing buffers. The dissociating properties of guanidine are superior to those of urea, but guanidine is considerably more expensive. The C-apoproteins have apparent molecular weights in urea solutions of about 25,000 (10). In guanidine HCl they behave as proteins of 6–10,000 MW. The chromatographic profile (Fig. 3) is similar, but not identical, to that obtained with SDS. The void volume peak contains apoB and some intermediate-molecular weight proteins (fraction 1, Fig. 3), even when media such as Bio-Gel A 5M or Sepharose 6B (13) are employed. ApoE and other large proteins elute over a broad area (fractions 2 and 3, Fig. 3) and apoproteins C-II and C-III in

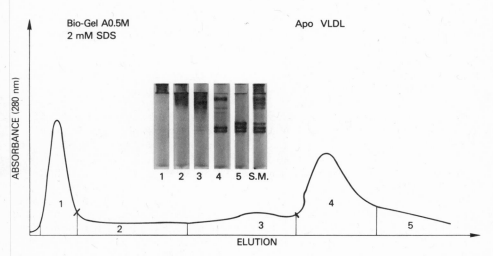

Fig. 2. Chromatography of apoVLDL on a 2.5 x 150 cm column of Bio-Gel A 0.5M. Conditions of polyacrylamide gel electrophoresis were as indicated in the legend to Fig. 1.

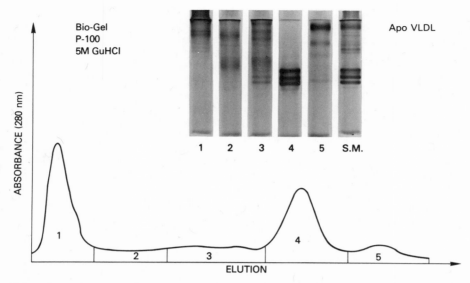

Fig. 3.   ApoVLDL fractionated on a 2.5 x 140 cm column of Bio-Gel
P-100 eluted with 0.1M NH$_4$HCO$_3$, 5M guanidine-HCl.

the second major peak (fraction 4, Fig. 3).   ApoC-I elutes in a
fairly well-defined peak (fraction 5, Fig. 3) after the other C-apoproteins.   The late elution of apoC-I is attributable to the dissociating power of guanidine and the low molecular weight of this
apoprotein ($\sim$ 7,000).

    Gel chromatography, therefore, permits rough quantitation of
the B and C apoproteins in VLDL.   The accuracy is perhaps better
when SDS buffers are used in apoB estimations and urea or guanidine-
HCl for the C-apoproteins.   The C-apoproteins can be further fractionated by ion-exchange chromatography on DEAE-cellulose.   DEAE-
Sephadex does not generate peaks as sharp or as reproducible as the
cellulose-based exchange medium.   Shore and Shore have applied whole
apoVLDL to DEAE-cellulose and resolved six to fifteen fractions (14,
15).   We have rarely been able to obtain homogeneous fractions, however, unless preliminary gel chromatography is used to separate the
C-apoproteins from the rest of the apoVLDL.   The proteins of intermediate molecular weight (fractions 2 and 3, Figs. 1-3) elute from
DEAE over a wide range of buffer conductivities and contaminate

several of the C-apoprotein fractions.   DEAE chromatography without
urea does not permit good resolution of the C-II and C-III apoproteins (11).   A gradient of Tris-HCl is better than one of NaCl.
While good separation of the C-apoproteins is possible (10), recov-
eries from DEAE are usually not greater than 75 or 80%.   It is not
known whether the losses are selective and DEAE chromatography has,
therefore, limited use in apoprotein quantitation.   Apoproteins B
and E are recovered in very low yields from DEAE and ion-exchange
chromatography is not suited to quantitation or purification of
these VLDL apoproteins.

ApoHDL.   Gel chromatography on columns eluted with 6-8M urea
have long been used to obtain estimates of the major protein components of apoHDL (7).   Chromatography in acetic acid has yielded
comparable results (8).   The major HDL apoproteins, apoA-I and apo-
A-II, elute in distinct but incompletely separated peaks (Fig. 4).
The void volume fraction (1, Fig. 4) is invariably present when
acetic acid is used for chromatography and is rarely seen when the
protein is solubilized and eluted with guanidine-containing buffers
(Fig. 5).   This peak consists mostly of irreversibly aggregated
protein (both apoA-I and apoA-II).   The first major peak (fraction
2, Fig. 4; and fraction 1, Fig. 5) contains the A-I apoprotein that

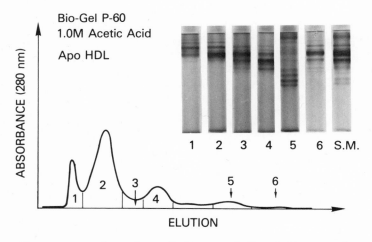

Fig. 4.   Chromatography of apoHDL on a 2.5 x 140 cm column of Bio-
Gel P-60 eluted with 1.0M acetic acid.

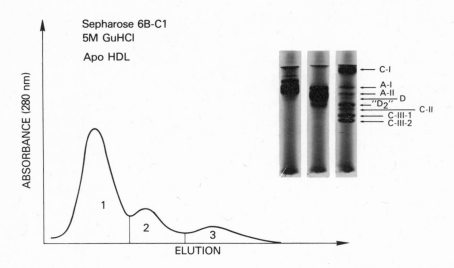

Fig. 5. Fractionation of apoHDL on a 2.5 x 140 cm column of Sepharose 6B-Cl eluted with 5M guanidine HCl.

usually is contaminated with apoA–II and occasionally with albumin and quantitatively minor HDL apoproteins. The valley between fractions 2 and 4 (Fig. 4) contains apoD (16, 17) and the second major peak (fraction 4, Fig. 4; and fraction 2, Fig. 5), the A–II apoprotein. The latter contains no tryptophan and consequently is not well–represented by the absorbance at 280 nm. The C–apoproteins elute next (fraction 5, Fig. 4; and fraction 3, Fig. 5). The C–apoprotein peak from HDL also contains the apoprotein recently described by Lim et al. (18) that has been tentatively designated "D–2". This latter apoprotein is apparent when most apoHDL samples are analyzed by polyacrylamide gel electrophoresis, but there is considerable variation among patients in the quantity present in apoHDL (cf., Figs. 4 and 5).

Several problems encountered in gel chromatography of apoHDL render it less than ideal for apoprotein quantitation. (1) Significant amounts of protein aggregate as a result of delipidation or incomplete dissociation during resolubilization. Protein aggregates elute in a void volume fraction that may account for variable (2–10%) proportions of the total recovered from chromatography. (2) In most, if not all, chromatographic systems, the A–I and A–II apoproteins are not completely separated. Moreover, both of the peaks comprising these apoproteins typically contain apoD. (3) ApoA–I often is detectable in all column fractions, particularly if buffer

systems without guanidine-HCl are used. This phenomenon is particularly troublesome if acetic acid is employed. A distinct peak containing apoA-I elutes even near the salt volume (fraction 6, Fig. 4)
where concentrated acetic acid used to solubilize the apoHDL emerges
from the column. The size of this peak increases with succeeding
column runs, suggesting resolubilization of precipitated apoA-I by
other HDL apoproteins and the concentrated acetic acid.

ApoHDL can also be fractionated by DEAE chromatography. Re-
coveries, however, are less than quantitative, the C-apoproteins
are not eluted in predictable positions, the A-I apoprotein is recovered in two or more peaks (19) over a broad range of conductivities (probably reflecting several states of aggregation), and the
purity of fractions is variable.

Quantitation by Polyacrylamide Gel Electrophoresis

Estimates of the apolipoprotein content of isolated lipoprotein
fractions can be obtained by polyacrylamide gel electrophoresis. The
values obtained in most circumstances are relative, not absolute. The
procedure can be employed to monitor changes in apoprotein content
with dietary or drug perturbations or it can be employed in semi-
quantitative comparisons of apoprotein patterns among different subjects.

The tetramethylurea (TMU) extraction procedure of Kane (6) has
been used by several laboratories in combination with photodensitom-
etry to analyze the protein content of apoVLDL and apoHDL. The
great advantage of the procedure resides in its application to small
quantities of lipoprotein. TMU appears to dissociate the lipoprotein lipid and selectively solubilizes all proteins except apoB (16).
It has been used for both estimations of relative proportions of apolipoproteins (20) and in efforts to absolutely quantify the apoprotein constituents of VLDL (21). The procedure has been described in
considerable detail (6, 21) and will not be recounted here. The use
of such assays for apoprotein quantitation demands extensive validation and rigid control of experimental conditions. Apoproteins in
pure form and in naturally occurring mixtures must be shown to produce a linear response in densitometric area. The limitations of
sensitivity must be defined and the identity (and homogeneity) of
stained bands confirmed by amino acid analysis and immunochemical
techniques. Comparison of mobilities of known apoproteins with
those in complex mixtures is not sufficient to establish identity.
When TMU is used for lipoprotein delipidation, the influence of
temperature on the recovery of soluble apoproteins must be considered (21). The volume of the stacking gels should be at least twice
that of the largest sample to insure that the proteins are well-
stacked before they enter the separating gel. ApoC-I does not stack
well in most alkaline gel systems because of its high isoelectric

point, and quantitation of this apoprotein is probably best performed in acid systems. Gel concentrations and buffer pH must be
standardized because of their obvious influence on R_f, zone width
and densitometric area. The disproportionate quantities of apoproteins in mixtures will often require electrophoresis at more than
one load to obtain reproducible quantitation. The absolute chromo-
genicities of different apoproteins are not the same, and must be
determined for the particular dye and staining conditions employed.

    The influence of sample aging on the polyacrylamide gel elec-
trophoresis pattern is one of the most difficult factors to control.
An extreme example of alteration of apoVLDL patterns is displayed in
Fig. 6. Comparison of the apoVLDL samples before fractionation
(S.M.) demonstrates that the intensity of the C-II band was greatly
diminished after storage of the plasma for one year at 4°C. The

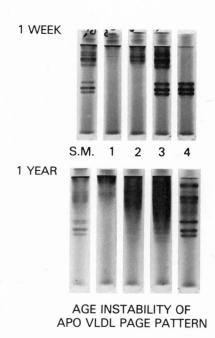

AGE INSTABILITY OF
APO VLDL PAGE PATTERN

Fig. 6. Alteration of the polyacrylamide gel electrophoresis patterns of apoVLDL and its gel chromatographic fractions after storage
of whole plasma for one year at 4°C. The apoVLDL shown in the upper
panel was fractionated on a column of Bio-Gel A 0.5M (Fig. 2), that
in the lower panel on Sepharose 6B-Cl eluted with 5M guanidine-HCl.

intermediate molecular weight apoproteins (fraction 2, Fig. 6) present distinct bands after one week of storage but a diffuse haze
after one year. This appearance is typical of denatured and aggregated proteins. Analysis of residual phospholipid in fraction 2
(Fig. 6) after one year shows quantities as high as 10-15% of the
protein weight after organic solvent extraction. It is possible
that peroxidation of unsaturated fatty acids and subsequent formation of covalent bands with the apoprotein accounts for this phenomenon. Inspection of fraction 4 (Fig. 6) shows that only a trace of
apoC-II band remains at one year, and there has appeared an intense
band, absent initially, with mobility distinctly less than apoC-II.
Kane and coworkers (21), moreover, have noted that the form of apo-
C-III of greatest mobility is diminished with storage, and we have
noted the appearance of several new, uncharacterized bands.

The capriciousness of column chromatography and polyacrylamide
gel electrophoresis in the quantitation of apoproteins in isolated
lipoproteins has been emphasized not to denigrate their usefulness,
but to underscore the limitations inherent in their use. They have
been widely employed in many laboratories including that of the
authors, but they will probably be displaced in time by immunochemical methods that are more readily standardized and which require
less sample manipulation.

TECHNIQUES APPLICABLE TO UNFRACTIONATED PLASMA

The specificity and sensitivity of several immunochemical
methods render them ideal for apolipoprotein quantitation in un-
fractionated plasma or serum in addition to isolated lipoprotein
fractions. All are based on the principle of titration - the reactivities of test samples are matched with those of known standards. The choice of techniques depends on several factors, including the concentration of the apoprotein in plasma (Table 1), the
number of samples to be processed, the degree of precision and
accuracy required, and the availability or expense of necessary
equipment and reagents. Pure antigens are mandatory for most assay
systems with the exception of two-dimensional immunoelectrophoresis
which will not be discussed. Monospecific antisera are necessary
for some, but not all, methods.

Purification of Antigens. The methods of apoprotein fractionation by gel and ion-exchange chromatography discussed above provide
general outlines of the most widely used techniques of apoprotein
purification. It is a common experience, however, that two chromatography steps are frequently inadequate to produce antigens of
sufficient homogeneity that truly monospecific antisera can be produced. We will briefly sketch here our own approaches for the
production of good antigens, recognizing that methods used by others
may be comparable or superior.

Table 1.   Approximate* Quantities of Apolipoproteins
           in Normal Plasma (mg/dl)

A-I	80-120
A-II	30-50
B	70-90
C-I	3-7
C-II	3-5
C-III	8-12
D	2-4
E	3-6
"D-2"	1-2

*Reliable estimates for several of these proteins
are not available. The values presented for
these are only educated guesses.

HDL to be used for apoA-I isolation is prepared by adjusting
fresh plasma to 1.090 g/ml with KBr and separating the VLDL, LDL,
and Lp(a) lipoproteins by ultracentrifugation. The infranatant
fraction is adjusted to 1.21 g/ml and the HDL isolated by a second
ultracentrifugation step. The density of the HDL is raised to 1.23
g/ml with solid KBr, an aqueous solution of KBr at 1.21 g/ml is
layered over it, and ultracentrifugation repeated to remove most of
the serum albumin. The HDL are dialyzed and totally delipidated
with the methanol-diethylether procedure described above. ApoHDL
is resolubilized in 5M guanidine HCl and fractionated on columns of
Bio-Gel A 0.5M, P-100 or Sephacryl. The latter may provide optimal
results. The several tubes around the peak of the apoA-I fraction
are pooled, dialyzed against 5mM NH_4HCO_3 and lyophilized. This crude
apoA-I is further purified by DEAE-cellulose chromatography in 6M
urea using a linear gradient from 0.05M to 0.15M Tris-HCl. The
apoA-I elutes in two or three poorly separated peaks, each of which
is individually pooled and analyzed for homogeneity by alkaline and
SDS polyacrylamide gel electrophoresis and amino acid analysis.
Only apoA-I that is pure by all criteria is used to raise antisera.

HDL for apoA-II isolation is prepared and delipidated as described for apoA-I. However, whole apoHDL is first applied to DEAE-
cellulose and the crude apoA-II fraction pooled, dialyzed and lyophilized. Final purification is achieved by chromatography in
guanidine HCl on columns of Bio-Gel P-100 or Sephacryl. If polyacrylamide gel electrophoresis or amino acid analysis suggest contamination, the apoA-II should be reduced, carboxymethylated and
subjected to a second gel chromatography step before use as antigen.

Antisera reactive with lipid-free B protein are not easily produced by immunization of animals with whole LDL. Therefore, apoB is

first produced in homogeneous form. LDL is isolated in a narrow
density range, 1.035 < d < 1.050 g/ml, from fresh plasma or sera.
The LDL is recentrifuged once at 1.050 g/ml, dialyzed, and delipidated with the methanol-diethylether solvents. Before drying under
nitrogen, solid SDS sufficient for a concentration of 0.2M in a protein solution of 10 mg/ml is added. The apoB is fractionated on a
column of Bio-Gel A 0.5M or A 1.5M eluted with 0.05M NH_4HCO_3, 2 mM
SDS. The tubes on either side of the void volume peak are pooled
and lyophilized, those on the ascending and descending limbs are
discarded. The SDS in the column effluent permits ready solubilization of the lyophilized apoB in 0.1M NH_4OH, and the resolubilized apoB is subjected to a second purification step on the same
column before use as antigen.

ApoC-I for immunization is obtained in the break-through volume
of the DEAE-cellulose columns used to fractionate the C-apoproteins
of apoVLDL (10). Before use as antigen, the apoC-I is purified by
gel chromatography in guanidine HCl on columns of Sephadex G-75,
Bio-Gel P-60 or Bio-Gel P-100. In spite of its low molecular weight,
apoC-I is very immunogenic.

ApoC-II is the most difficult of the C-apoproteins to obtain in
a high degree of purity. Chromatography of apoVLDL on Bio-Gel A
0.5M (Fig. 2) is most useful for this purpose because apoC-II is recovered near the end of the chromatogram free of contamination with
intermediate-molecular weight proteins. When apoC-II is to be used
for immunization, the appropriate peak from DEAE-cellulose (10)
should be cut very narrowly to minimize the possibility of contamination with apoC-III. The latter cannot be separated from apoC-II
by gel chromatography since their molecular weights are almost identical. Antisera raised to apoC-II that react with apoC-III can be
easily adsorbed, since apoC-III is easy to prepare in homogeneous
form.

ApoC-III for immunization is obtained as described (10). Apo-
C-III-1 is best for this purpose, and only peak tubes from DEAE-
cellulose (10) should be used as antigen. The apoC-III-2 peak is
frequently contaminated with proteins of similar electrophoretic
mobility and size. Antisera raised to apoC-III-2 are often oligo-
specific.

The authors have not succeeded in purifying apoD. Published
methods for apoD isolation (16, 17) detail the difficulties usually
encountered. Fortunately, many antisera raised to crude apoA-II
react with apoD and specific anti-apoD can be obtained by adsorption.

ApoE, like apoD, is purified with considerable difficulty. A
crude preparation is obtained by chromatography of apoVLDL (ideally
from a patient with type 3 hyperlipoproteinemia) on Bio-Gel A 0.5M
(SDS buffer system) and rechromatography on Sephacryl or Bio-Gel A

0.5M or 1.5M eluted with 5M guanidine HCl. Preparative SDS polyacrylamide gel electrophoresis is the final purification step. This
can be accomplished using a commercial preparative gel apparatus
(22). Alternatively, the apoprotein E zone (identified by use of a
fixed and stained marker gel) can be sliced from analytical calibre
gels, the gels homogenized in 0.1M NH$_4$OH, and the gel slurry separated by low speed centrifugation. If the slurry is extracted a
second time with 0.1M NH$_4$HCO$_3$, about 60% of the apoE can be recovered. After dialysis and lyophilization, a preparation suitable for
immunization is obtained. Alternatively, the minced gel slices can
be used directly for immunization.

     Preparation of Antisera. One to three mg of pure apoprotein is
usually sufficient to raise moderate titer antisera in rabbits or
other small rodents. Useful antisera have been obtained, however,
with a total of 0.5 mg protein. Larger quantities, but not more
than 6 or 8 mg, are sometimes needed for sheep or goats. Rabbits
can be injected in foot pads or (more humanely) intradermally on the
back. The apoprotein solution is thoroughly emulsified in complete
Freund's adjuvant. One-fourth of the total available dose is administered each injection. A minimum of three injections at bi-
weekly intervals is necessary and higher-titer antisera can be produced if sufficient material is available for 4-6 injections. The
use of larger amounts of protein at less frequent intervals usually
induces lower titer antisera. If the immunochemical assay is to be
used for a considerable period of time, it is useful to pool several
bleeds from rodents or at least two from sheep or goats to avoid repeated tedious characterization and standardization of antisera.

     It is often desirable to optically clarify antisera or to re-
move plasma proteins that interfere with immunochemical assays.
Antisera produced in rabbits may be grossly lipemic, a reflection of
even mild infection at injection sites. Lipoproteins can be separated from the gammaglobulins and other serum proteins by a single
ultracentrifugation at 1.21 g/ml. The infranatant fraction is
dialyzed against 0.2M NaCl, 0.05M NH$_4$HCO$_3$, filtered through glass
wool, bacteriostatic agents are added, and the antisera frozen until
use. Hemolyzed antisera raised in sheep can be clarified and the
gammaglobulin fraction isolated by DEAE-cellulose chromatography.
The column is equilibrated with 0.03M Tris-HCl, pH 8.0, and the antisera dialyzed against the same buffer. Hemoglobin and most of the
plasma proteins are bound to the DEAE and the gammaglobulins can be
eluted with the equilibration buffer and lyophilized. Alternatively,
the gammaglobulins in crude form can be precipitated from the antisera by mixing with an equal volume of 28% (w/v) Na$_2$SO$_4$, and washing the precipitate with 14% Na$_2$SO$_4$. Lipoproteins are precipitated
under these conditions, producing turbidity in the resolubilized
gammaglobulin solution. A simple method avoiding precipitation of
lipoproteins is that employing caprylic acid for isolation of
mammalian IgG (24).

Antisera or purified gammaglobulins are stored frozen in small aliquots to avoid repeated freezing and thawing. A satisfactory bacteriostatic agent is a mixture of 3.45×10^{-3}M NaN$_3$ and 1.5×10^{-4}M thimerosal (1:1,000, v:v). Antisera stored at 4°C for prolonged periods usually lose activity.

Gel Immunodiffusion

Quantitation of plasma proteins by a variety of immunodiffusion techniques has been used for 20 years. Agar and agarose have widely been chosen as the stabilizing media, although cellulose acetate polyacrylamide, filter paper, and other media have found occasional use. The liquid phase usually contains 0.15 - 0.20M NaCl and an appropriate buffer to maintain pH at 7.0-8.5. At pH > 9.0, antigen-antibody complexes may be dissolved and many antigens are insoluble at more acid pH. When chicken antibodies are used, a gel containing 10% NaCl has been recommended. We routinely use a 0.05M barbital (Veronal) buffer of pH 8.2 and an agar or agarose final concentration of 1%. In preparation for quantitative immunodiffusion, it is convenient to prepare a 2% gel solution by boiling the agar or agarose in the liquid phase containing 1:10,000 thimerosal (w:v), filtering the solution through a hot glass funnel with cheese cloth or glass wool, and storing the gel at 4°C in 25 ml or 50 ml Ehrlenmeyer flasks covered with parafilm. Plates for single immunodiffusion are prepared by melting the gel in a boiling water bath. The gel is cooled to 56°C (before mixing with an equal volume of antiserum), diluted in the liquid phase, and heated to 56°C. A 1-2 mm layer is poured on a glass plate or petri dish (23). The dilution of antiserum that is appropriate must be determined empirically and is highly contingent on the antibody titer. Dilutions in a range of 1:10 to 1:1,000 may be appropriate. When only small dilutions are possible, the procedure may prove too expensive in terms of precious antisera. Techniques are available, however, for increasing the sensitivity (24). The titers of antisera raised to many of the apolipoproteins are often much lower than those of commercial antisera raised to abundant proteins. This is usually secondary to the limited quantities of homogeneous antigens available for injection and the impossibility of immunizing several animals simultaneously.

Specific antisera are generally required to use the single radial immunodiffusion (RID) method for protein concentration. Polyspecific antisera will generate multiple precipitin rings, and the correlation of individual precipitin rings with specific antigens can be a cumbersome undertaking (25). Dilutions of antigen of known concentration are used to construct the standard curve, and each point is determined in duplicate on different areas of the plate. In addition, a standard preparation of lipoprotein solution, serum or plasma should be used with every assay to control for interassay variation.

The preparation of test samples for analysis requires special attention, particularly when purified apoproteins serve as standards and lipoproteins or serum are to be assayed. The reactivity of antigens that are complexed in one or more lipoproteins should be considered unknown, and a variety of pretreatment procedures evaluated. Total delipidation of lipoproteins or plasma with organic solvents is often considered optimal to insure maximum reactivity. Treatment with organic solvents, however, is time-consuming and complicated by uncertainty concerning selective or non-specific losses, and alteration of physical state (e.g., aggregation) and thereby immunoreactivity of the apoproteins. The use of ionic or non-ionic detergents, or dissociating agents such as guanidine and urea, to quantitatively unmask antigenic sites has been disappointing. Albers and coworkers (26) have treated plasma and HDL samples with TMU and a Tris buffer containing 8M urea, and have apparently achieved adequate exposure of the antigenic sites of apoA-I. TMU treatment, therefore, appears promising as an agent to maximize immunoreactivity of apolipoproteins and to minimize the handling of samples before actual assay.

The theoretical basis of quantitation in RID is still not well-defined. In general, the diameter of the precipitin ring is directly related to the quantity of antigen in the well and inversely related to the concentration of antibody in the gel. The reaction cannot be viewed as proceeding at equivalence, however, a fact visually confirmed by the non-uniformity of precipitate within the precipitin area (28). The usual response variables for RID have been the area, diameter, or square of diameter of the precipitin ring. A linear relationship of dose to squared diameter (29) or area (30) is expected. The appropriate duration of incubation before measurement of the precipitin ring is determined empirically. The choice is dictated by considerations of convenience as well as the characteristics of the systems. Readings may be taken as early as 24 hours in some cases (29), although the reaction may proceed for 14 days or more (27). Impedence to diffusion of large macromolecules is known to artifactually reduce their apparent concentrations (28), a phenomenon that may limit the potential usefulness of RID for apoB quantitation. Similarly, apoproteins bound to lipoproteins cannot be expected to behave identically to their lipid-free counterparts.

The limit of sensitivity of most RID systems is on the order of 1 mg/dl. RID is theoretically well-suited for the quantitation of apolipoproteins A-I, A-II and B. ApoD, apoE and the C-apoproteins are present in undiluted plasma at concentrations near the practical lower limit of RID analyses. Minor modifications of the classical method (24), however, may even permit their quantitation.

RID, like all immunoassays, requires very careful attention to technical details and quality control. Errors can arise and be compounded by inaccurate dilution or application of test samples, variations of gel thickness within a plate, incomplete mixing of antiserum and gel, inadequate spacing of test wells, dehydration of gel, etc. Nevertheless, the method requires a minimum amount of specialized equipment and technical expertise, reagents are inexpensive, and a large number of samples can be economically processed.

Electroimmunoassay

Two quantitative immunoelectrophoretic methods have been applied to the study of apolipoproteins. Bidimensional crossed immunoelectrophoresis potentially can be used with polyspecific antisera and may permit simultaneous quantitation of more than one antigen in a complex mixture. Its use is being evaluated in several laboratories, but as yet there are no published examples of its application to apolipoprotein quantitation. The technique is technically demanding and not readily applied to large numbers of samples. A useful literature review of this and other immunoelectrophoretic methods has been published by Verbruggen (31) and this should be consulted for a general summary of theoretical and practical considerations.

Electroimmunoassay (EIA) has been used to quantitate the proteins of both alpha- (32-35) and betalipoproteins (36-38). A monospecific antisera is incorporated in an agarose gel as described above for RID. Antigen is added to wells punched in the gel and electrophoresis results in the production of rocket or spiked shaped precipitins, the height or area of which is proportional to antigen concentration (39, 40). The conditions of electrophoresis are chosen so that antibody is presumed to be stationary (31). The anodally migrating antigens precipitate with antibodies but unbound antigen within the precipitin cones migrates into and continues to redissolve the precipitate, displacing the precipitin zones toward the anode, until free antigen is finally exhausted.

In EIA, as in RID, the assay conditions must be determined empirically. Conditions are sought wherein the smallest amount of antigen in the lowest possible antiserum concentration gives a well-defined precipitin peak of 2-5 cm in height. The useful antiserum concentration may be as low as 0.5% or as high as 20%. The minimum amount of antigen that can be confidently quantitated may range from 3-100 ng. The incorporation in the gel of so-called "enhancers" of immune precipitation, such as dextrans or polyethylene glycol, can augment the sensitivity by a factor of 2-5 (41, 42). Using Dextran T10 (Pharmacia Fine Chemicals), Curry et al. were able to measure 20 ng of apoA-I or apoA-II in EIA.

The factors which can affect quantitation with EIA and their

evaluation are discussed elsewhere (31) and will not be elaborated
here. The results to date with apolipoproteins have been puzzling
and at variance with much other published work. Durrington and coworkers (38) discovered that response curves of LDL and serum were
not parallel in EIA and a secondary standardization was necessary
to obtain quantitation. Farish et al. (34) obtained values for
"apolipoprotein A" that are higher than those predicted from ultracentrifugation (1) or radioimmunoassay (vide infra). Still higher
"apoA" levels were reported by Curry et al. (35), who, moreover,
found apoA-I/apoA-II weight ratios considerably different from those
estimated by other techniques. In neither of these studies was it
deemed necessary to delipidate the plasma or lipoprotein solutions
to unmask HDL antigenic sites, whereas such treatment has generally
been required in other immunoassays. In addition, the possibility
that isolated apolipoprotein "standards" were less reactive than the
naturally occurring species in intact lipoproteins was not excluded.
Both studies suggested that "apoA" could be quantitated directly by
EIA using a polyspecific antiserum to HDL. This impression is con-
ceptually misleading. The antisera employed undoubtedly contained
antibodies recognizing both the A-I and A-II apoproteins, but the
EIA theory would predict that only one group of antibodies (probably
that present in highest titer) determined the EIA result. The "apo-
A" quantitation obtained, therefore, would be valid only if the test
sample had the same (A-I/A-II) and (A-I+A-II/total HDL apoprotein)
ratios as the "standard" apoHDL. The results presented in one paper
(35) suggest that such an assumption may not be justified.

EIA has been recommended for protein quantitation because of
its high sensitivity, conservation of antisera, and rapidity on
comparison with diffusion techniques. However, it is not a simple
method to execute, standardize, or quality control, and it is not
applicable to large scale studies. Extensive validation is mandatory before quantitation with this method can be accepted without
qualification.

Radioimmunoassay

The extreme sensitivity of radioimmunoassay (RIA) is not required for quantitation of most of the known plasma apolipoproteins
(see Table), but it is ideal for studies of apoprotein metabolism
in cell suspensions, tissue culture and isolated perfused organs.
The high degree of automation possible with this technique, moreover, renders it particularly suitable for the analysis of large
numbers of samples. Of the techniques discussed above, only radial
immunodiffusion can be comparably adapted to mass screening efforts.

The basic elements of RIA include: (1) antibody; (2) labeled
and unlabeled antigen; (3) test sample; (4) incubation; (5) separation of antibody-bound and free antigen; (6) radioactivity measure-

ment; and (7) data reduction. Items four through seven present no problems peculiar to apolipoproteins and are treated in detail in most monographs and reviews of RIA. The first three elements warrant consideration here for reasons apparent in the discussion of other immunoassays.

Antisera used in RIA generally need not be specific for a single antigen except in the unusual circumstance where the antigen to be labeled is not pure. Grossly impure antigens are of little value in any event because of inevitable difficulties in establishing the immunoreactivity of the antigen after radiolabeling. It is desirable that the antibodies be directed against regions of the apolipoprotein molecule that are fully exposed in native lipoproteins. This circumstance would minimize or obviate the delipidation procedures necessary before actual assay. Theoretically, such antisera might be produced if native lipoproteins are employed for immunization. Most of the RIAs that have been developed for apoB have employed antisera raised to LDL (43-46) and there has been no need to treat test samples with detergents or organic solvents. However, apoB may represent a special case. An RIA based on an antiserum to lipid-free and chromatographically purified apoB has been reported (47). Pretreatment of test samples was unnecessary, but apoB in delipidated lipoproteins was only about 15% as reactive as native LDL (47). The standard antigen in all these assays was LDL, not apoB, and the poor reactivity of lipid-free B protein was ascribed without proof to aggregation (47). There exists additional uncertainty concerning the potential of RIA with LDL as standard to quantitate the apoB content of VLDL (44, 45, 47).

The possibility of using antisera to native lipoproteins to quantitate other apolipoproteins has not been fully evaluated. The antisera employed for RIA of apoA-I (48-51) and apoA-II (51, 52) have been raised to purified apoproteins with the exception of that reported by Fainaru et al. (49) in which antisera to both HDL and apoA-I were used. In all assays except the latter, pretreatment of test samples was required. Schonfeld and Pfleger (48) found that less than 30% of plasma apoA-I was recognized in RIA without organic solvent delipidation. Treatment with urea and guanidine, or freeze-thawing produced suboptimal quantitation. Karlin and coworkers (50) were able to expose the apoA-I antigenic sites in plasma by the simple maneuver of diluting and heating. Only Fainaru et al. (49) have claimed that plasma apoA-I could be quantitated without pretreatment of test samples, and this was possible in their assay system using antibodies raised to HDL, apoHDL, and apoA-I. It is unlikely, therefore, that the specificities of the antisera account for the differences between the assay of Fainaru (49) and those of Schonfeld and Karlin. Differences in antibody specificity, however, may account for the relative ease with which Karlin et al. exposed the immunoreactive sites of apoA-I in their assay (50).

Schonfeld et al. (53) have demonstrated that an anti-apoA-I serum
in their laboratory contains at least two subpopulations of antibodies specific for the carboxyl and amino-terminal of apoA-I and
without cross-reactivity. One of these antibody subpopulations was
more reactive with native HDL (53).

   If VLDL or HDL are used to produce antisera to their apolipoproteins, it will probably be necessary to purify the antisera by
immunoadsorption before use in RIA. This is because of the apolipoprotein heterogeneity of these classes of lipoproteins and the
rapidity with which these lipoproteins exchange. Thus, if an antiserum used in an RIA for apoC-III also precipitated apoB containing
lipoproteins, then the presence of C-III and B on the same lipoprotein could lead to "nonspecific" precipitation of C-III by anti-
apoB.

   Radiolabeling of apoproteins has generally been performed
chemically with chloramine-T (54) or iodine monochloride (55), or
enzymatically with lactoperoxidase (56). The chloramine-T procedure
has been satisfactory for isolated apoproteins, but one of the
other procedures is preferable when labeled lipoproteins are to be
used as standards. Lipoproteins from animals particularly may
contain an abundance of polyunsaturated fatty acids and a large
amount of label may be incorporated in the lipid when chloramine-T
is used (57).

   The high degree of sample dilution necessary to conserve antisera in the RIA of apolipoproteins is one of the major drawbacks of
this method. Automation of the procedure is also relatively expensive and the work of pretreating samples is tedious and time-
consuming. (Attempts to use TMU for this purpose have not been
published). The sensitivity and accuracy of the technique, however, hold great appeal in innumerable research situations and
increasing use of RIA and other immunoassays can be anticipated in
studies of lipoprotein physiology.

REFERENCES

1. Havel, R.J., Eder, H.A., and Bragdon, S.H. J. Clin. Invest.
    34, 1345-1353, 1955
2. Herbert, P.N., Forte, T.M., Shulman, R.S., LaPiana, M.J., Gong,
    E.L., Levy, R.I., Fredrickson, D.S., and Nichols, A.V. Prep.
    Biochem. 5, 93-129, 1975
3. Scanu, A.M., and Edelstein,C. Anal. Biochem. 44, 576-588, 1971.
4. Folch, J., Lees, M., and Sloane-Stanley, G.H. J. Biol. Chem.
    226, 497-509, 1957
5. Helenius, A., and Simons, K. Biochem. 10, 2542-2547, 1971.
6. Kane, J.P. Anal. Biochem. 53, 350-364, 1973
7. Scanu, A., Toth, J., Edelstein, C., Koga, S., and Stiller, E.

Biochem. 8, 3309–3316, 1969

8. Rudman, D., Garcia, L.A., and Howard, C.H. J. Clin. Invest. 49, 365–372, 1970

9. Scanu, A.M., Lim, C.T., and Edelstein, C. J. Biol. Chem. 247, 5850–5855, 1972

10. Herbert, P.N., Shulman, R.S., Levy, R.I., and Fredrickson, D.S. J. Biol. Chem. 248, 4941–4946, 1973

11. Brown, W.V., Levy, R.I., and Fredrickson, D.S. J. Biol. Chem. 244, 5687–5694, 1969

12. Reisfeld, R.A., and Small, P.A., Jr. Science 152, 1253–1255, 1966

13. Shelburne, F.A., and Quarfordt, S.H. J. Biol. Chem. 249, 1428–1433, 1974

14. Shore, B., and Shore, V. Biochem. 8, 4510–4516, 1969

15. Shore, V.G., and Shore, B. Biochem. 12, 502–507, 1973

16. McConathy, W.J., and Alaupovic, P. FEBS Letts. 37, 178–182, 1973

17. Kostner, G.M. Biochim. Biophys. Acta 336, 383–395, 1974

18. Lim, C.T., Chung, J., Kayden, H.J., and Scanu, A.M. Biochim. Biophys. Acta 420, 332–341, 1976

19. Lux, S. E., and John, K. M. Biochim. Biophys. Acta 278, 266–270, 1972

20. Carlson, L.A., and Ballantyne, D. Atherosclerosis 23, 563–568, 1976

21. Kane, J.P., Sata, T., Hamilton, R.L., and Havel, R.J. J. Clin. Invest. 56, 1622–1634, 1975

22. Utermann, G. Hoppe-Seyler's Z. Physiol. Chem. 356, 1113–1121, 1975

23. Steinbuch, M., and Andran, R. Arch. Biochem. Biophys. 134, 279–284, 1969

24. Albers, J.J., and Hazzard, W.R. Lipids 9, 15–26, 1974

25. Darcy, D.A. Clin. Chim. Acta 38, 329–337, 1972

26. Mancini, G., Nash, D.R., and Heremans, J.F. Immunochem. 7, 261–264, 1970

27. Albers, J.J., Wahl, P.W., Cabana, V.G., Hazzard, W.R., and Hoover, J.J. Metabol. 25, 633–644, 1976

28. Trautman, R., Cowan, K.M., and Wagner, G.G. Immunochem. 8, 901–916, 1971

29. Fahey, J.L., and McKelvey, E.M. J. Immunol. 94, 84–90, 1965

30. Mancini, G., Carbonara, A.O., and Heremans, J.F. Immunochem. 2, 235–254, 1965

31. Verbruggen, R. Clin. Chem. 21, 5–43, 1975

32. Johansson, B.G., and Laurall, C.-B. Scand. J. Clin. Lab. Invest. 23, 231–233, 1969

33. Johansson, B.G., and Medhus, A. Acta Med. Scand. 195, 273–277, 1974

34. Farish, E., Shepherd, J., Lawrie, T.D.V., and Morgan, H.G. Clin. Chim. Acta 62, 97–101, 1975

35. Curry, M.D., Alaupovic, P., and Suenram, C.A. Clin. Chem. 22, 315–322, 1976

36. Kahan, J., and Sundblad, L.   Scand. J. Clin. Lab. Invest. 24,
     61-68, 1969
37. Lasser, N. L., Roheim, P.S., Edelstein, D., and Eder, H.A.   J.
     Lipid Res. 14, 1-8, 1973
38. Durrington, P.N., Whicher, J.T., Warren, C., Bolton, C.H., and
     Hartog, M.   Clin. Chim. Acta 71, 95-108, 1976
39. Laurell, C.-B.   Anal., Biochem. 15, 45-52, 1966
40. Laurell, C.-B.   Scand. J. Clin. Lab. Invest. 29, Suppl. 124, 21-
     37, 1972
41. Harrington, J.C., Fenton, J.W., II, and Pert, J.H.   Immunochem. 8,
     413-421, 1971
42. Kostner, G., and Holasek, A.   Anal. Biochem. 46, 680-683, 1972
43. Eaton, R.P., and Kipnis, D.M.   J. Clin. Invest. 48, 1387-1396,
     1969
44. Schonfeld, G., Lees, R.S., George, P.K., and Pfleger, B. J. Clin.
     Invest. 53, 1458-1467, 1974
45. Albers, J.J., Cabana, V.G., and Hazzard, W.R.   Metab. 24, 1339-
     1351, 1975
46. Bautovich, G.J., Simons, L.A., Williams, P.F., and Turtle, J.R.
     Atherosclerosis 21, 217-234, 1975
47. Thompson, G.R., Birnbaumer, M.E., Levy, R.I., and Gotto, A.M., Jr.
     Atherosclerosis 24, 107-118, 1976
48. Schonfeld, G., and Pfleger, B. J. Clin. Invest. 54, 236-246, 1974
49. Fainaru, M., Glangeaud, M.C., and Eisenberg, S. Biochim. Biophys.
     Acta 386, 432-443, 1975
50. Karlin, J.B., Juhn, D.J., Starr, J.I., Scanu, A.M., and Rubenstein,
     A.H.   J. Lipid Res. 17, 30-37, 1976
51. Assmann, G., Smootz, E., Adler, K., Capurso, A., and Oette, K.
     J. Clin. Invest. 59, 565-575, 1977
52. Mao, S.J.T., Gotto, A.M., Jr., and Jackson, R.L.   Biochem. 14,
     4127-4131, 1975
53. Schonfeld, G., Bradshaw, R.A., and Chen, J.-S.   J. Biol. Chem.
     251, 3921-3926, 1976
54. Roth, J.   In O'Malley, B.W.O., and Hardman, J.G. (Eds.): Methods
     in Enzymology XXXVII.   New York, Academic Press, 1975, pp. 223-233
55. McFarlane, A.S., Munro, H.N., and Allison, J.B. (Eds.: Meta-
     bolism of Plasma Proteins in Mammalian Protein Metabolism.
     Academic Press, New York, 1964, p. 331.
56. Marchalonis, J.J.   Biochem. J. 113, 294-305, 1969
57. Fidge, N.H., and Poulis, P.   Clin. Chim. Acta 52, 15-26, 1974

ISOLATION AND COMPOSITION OF HUMAN PLASMA APOLIPOPROTEINS

Henry J. Pownall

Baylor College of Medicine and the Methodist Hospital

Houston, Texas 77030, U.S.A.

I. INTRODUCTION

The biochemist's attempts to understand the structure and function of the human plasma lipoproteins have expanded greatly in recent years due to their possible role in the development of atherosclerosis. Five of the major apolipoproteins have been sequenced and our view of the structural organization of lipoproteins has seen considerable refinement. All of these achievements depend, in one way or another, upon the investigators skill in isolating both intact lipoproteins and apolipoproteins in a highly purified and homogeneous state. For this reason, we present in this chapter a description of some of the tried and proven methods for the purification of the lipoproteins and their component apolipoproteins and lipids. This presentation will not include the purification of abberant lipoproteins; the less frequently used or exotic techniques will be mentioned only to provide a reference for those who may be interested in the more specialized methods. Three reviews of lipoproteins provide ample source material on those areas of lipoprotein structure and function which are outside the scope of this volume (1-3).

II. COMPOSITION OF THE HUMAN PLASMA LIPOPROTEINS

A. Composition of Human Plasma High Density Lipoprotein (HDL)

The lipid-protein of HDL is depicted in Figure 1. The HDL (HDL_2 and HDL_3) are approximately 50% protein and 50% lipid. The protein composition is predominantly apoA-I (71%) and apoA-II (21%) with small amounts (8%) of apoC and apoD composing the remainder (4). The ratio of apoA-I to apoA-II was reported as 2:1 (5). The sequence of amino acids composing apoA-I and apoA-II are known (6,7) and are given in Figure 2 in terms of the one letter symbols for the amino acids (8). Phospholipids and cholesterol represent the major lipids of HDL; smaller amounts of cholesteryl esters and mono-, di-, and triglycerides account for the remainder. Phosphatidylcholine (75%) and sphingomyelin (13%) represent the dominant phospholipids which are accompanied by smaller quantities of other phospholipids and lysophosphatides (4).

The fatty acid composition of the individual lipids of the HDL from normal fasting subjects on an ad lib diet are shown in Table 1. Oleate, linoleate and palmitate represent the major fatty acids of the phospholipids and triglycerides whereas linoleate is the predominant fatty acid of the cholesteryl esters (9).

B. Composition of Human Plasma Low Density Lipoprotein (LDL)

The composition of plasma LDL from normal subjects is given in Figure 3. The particle contains 25% protein and 75% lipid. The major protein of LDL, apoB, represents its entire protein content. LDL contains 37% cholesteryl ester, 21% phospholipid, 9% cholesterol and 8% triglyceride. The major phospholipid classes are phosphatidylcholine (65%) and sphingomyelin (25%). The fatty acid composition of the individual lipid classes of LDL isolated from normal fasted subjects are given in Table 2; palmitate and oleate are the major fatty acids in the phospholipids of LDL (4). Cholesteryl esters of LDL are composed mainly of oleate and linoleate whereas the LDL triglycerides are composed largely of palmitate and oleate (4).

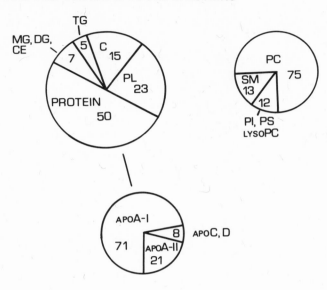

Figure 1: Lipid-protein composition of HDL.

apoA-I	DEPPQ SPWDR VKDLA RVYVD VLKDS GRDYV	30
	SQFQG SALGK QLNLK LLWDD VTSTF SKLRQ	60
	ELGPV TEEWF NDLQE KLNLE KETGE LRQEM	90
	SKDLE EVKAK VQPYL DDFQK KWQEM ELYRQ	120
	KEVPL RAELQ EGARQ KLHEL QEKLS PLGEE	150
	MRDRA RAHVD ALRTH LAPYS DELRQ RLAAR	180
	LEALK ENGAG RLAEY HAKAT EHLST LSEAK	210
	KPALE DLRQG LLPVL ESFKV SFLSA LEEYT	245
	KLNTQ	
apoA-II	ZAKEP CVESL VSQYF QTVTD YGKDL MEKVK	30
	SPELQ AQAKS YFEKS KEQLT PLIKK AGTEL	60
	VNFLS YFVEL GTQPA TQ	77

Figure 2: Amino acid sequences of apoA-I and apoA-II monomer. The amino terminal residue of apoA-II is pyrrolidone carboxylic acid. ApoA-II occurs as a dimer of identical chains linked by a disulfide bridge at residue-6.

Table 1

Average Fatty Acid Composition of HDL$_3$ of Four Normal
Subjects on an <u>Ad</u> <u>Lib</u> Diet

Fatty acid[a]	Phospholipids	Cholesteryl esters	Triglycerides
14:0	1.02	1.60	2.79
16:0	30.3	13.5	26.4
18:0	17.7	1.46	5.46
16:1	1.20	2.99	5.01
18:1	14.4	19.1	39.5
18:2	24.5	56.9	19.2
18:3,20:3	4.11	trace	trace
20:4	9.56	4.56	trace

——— a
        Adapted from Morrisett el al. (10). X:Y denotes X carbons
in an acyl chain having Y double bonds.

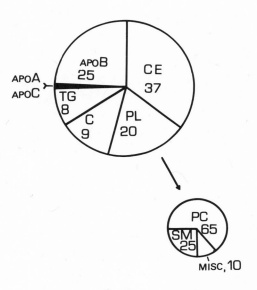

Figure 3: Lipid-protein composition of LDL.

Table 2

Average Fatty Acid Composition of LDL of Four Normal
Subjects on an Ad Lib Diet

Fatty acid[a]	Phospholipids	Cholesteryl esters	Triglycerides
14:0	0.57	1.04	2.24
16:0	31.2	14.8	28.6
18:0	18.4	0.96	4.53
16:1	trace	2.53	3.87
18:1	14.6	21.4	41.7
18:2	25.5	56.6	17.6
18:3,20:3	3.64	trace	trace
20:4	7.09	trace	trace

———— a

Same footnotes as Table 1.

C. Composition of VLDL

Figure 4 gives the lipid and protein composition of VLDL,
the lipoprotein responsible for the transport of endogenous
triglyceride (4). Approximately 10% of the particle is protein;
these are apoB (37%), apoC-I (4%), apoC-II (6%) and apoC-III,
and apoC-III$_2$ (40%). The latter differ in their respective
contents of 1 or 2 moles of sialic acid. All three of the C
proteins have been sequenced (10-13) and their primary structures are given in Figure 5.

The lipid composition of VLDL reflects its role as a
vehicle for triglyceride transport. The triglycerides compose
59% of the particle weight (4). The remainder is made up of
cholesterol (10%), cholesteryl esters (5%), and phospholipid
(15%). The fatty acid composition of each lipid class of VLDL
appears in Table 3. Palmitate is the major fatty acid of the
VLDL phospholipids with approximately equal amounts of linoleate, oleate and stearate. The VLDL cholesteryl esters are
mainly linoleate, oleate and palmitate whereas the VLDL triglycerides are largely palmitate and oleate (9).

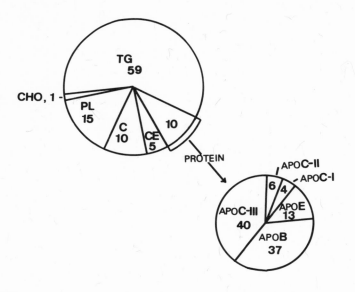

Figure 4: Lipid-protein composition of VLDL.

apoC-I	TPDVS SALDK LKEFG NTLED	20
	KAREL ISRIK QSELS AKMRE	40
	WFSET FQKVK EKLKI DS	57

apoC-II	TEQPQ QDEMP SPTFL TEUKE	20
	WLSSY QSAKT AAQNL YEKTY	40
	LPAVD EKLRD LYSKS TAAMS	60
	TYTGI FTDQV LSVLK GEE	78

apoC-III	SEAED ASLLS FMQGY MKHAT	20
	KTAKD ALSSV QSQQV AAQQR	40
	GWVTD GFSSL KDYWS TVKDK	60
	FSEFW DLDPE VRPTS AVAA	79

Figure 5: The amino acid sequence of the C-proteins, apoC-I, apoC-II and apoC-III. Threonine-74 in apoC-III is covalently bonded to one molecule of galactosamine and galactose and 0, 1 or 2 molecules of sialic acid (designated C-III$_0$, C-III$_1$ or C-III$_2$).

Table 3

Average Fatty Acid Composition of the VLDL from Four Normal Subjects on an Ad Lib Diet

Fatty acid[a]	Phospholipids	Cholesteryl esters	Triglycerides
14:0	2.99	6.19	3.40
16:0	35.3	21.2	31.8
18:0	20.4	6.25	4.57
16:1	trace	3.16	5.02
18:1	17.5	23.7	41.7
18:2	20.6	35.2	13.3
18:3, 20:3	trace	trace	trace
20:4	4.85	trace	trace

——— a

Same as Table 1.

III. ISOLATION OF HUMAN PLASMA LIPOPROTEINS

After addition of 0.01% azide the HDL, LDL and VLDL may be isolated according to their densities using the sequential flotation technique of Scanu (15). Plasma is adjusted to a background density of 1.006 g/ml by the addition of KBr and centrifuged in polyalamar tubes using an anglehead rotor (e.g. Beckman 60 Ti) for 12 hrs. at 55,000 rpm. This floats the VLDL and, if any are present, the chylomicrons; these are removed by tube slicing. The infranatant density is readjusted to d= 1.063 g/ml and recentrifuged for 24 hrs. in a 60 Ti at 55,000 rpm. The floated LDL are again removed by tube slicing. The combined infranatant density is adjusted to d=1.21 g/ml and centrifuged for 48 hrs in a 60 Ti rotor at 55,000 rpm. The resulting supernatant contains HDL and frequently albumin. The latter may be removed by a second 48 hr. centrifugation in which the supernatant is placed in the lower half of the cen-trifuge tube and the upper half filled with a KBr solution of d=1.21 g/ml. The floated HDL may be removed by tube slicing and infranatant discarded.

Table 4

Properties of the Human Plasma Lipoproteins[a]

	Particle radius, nm	Density, g/ml	Density range for isolation, g/ml
Chylomicrons	35.0	0.93	1.006
VLDL	12-35	0.97	1.006
LDL$_1$	11-12	1.003	1.006-1.019
LDL$_2$	10-11.5	1.034	1.019-1.063
HDL$_1$	-	1.05	-
HDL$_2$	3.5-5	1.094	1.063-1.125
HDL$_3$	2-3.5	1.145	1.125-1.210
VHDL	-	1.25	1.210-1.250

[a]Adapted from Scanu et al. and ref. cited therein (2).

The lipoproteins also exist as subclasses based on their densities. These may be isolated by a similar procedure using different densities to recentrifuge a given lipoprotein class. The density ranges for the various lipoproteins are given in Table 4.

An alternative procedure proposed by Rudel et al. (15) is based on the large differences in the sizes of HDL, LDL and VLDL given in Table 4. All of the LPs are first concentrated by a 24 hr. centrifugation in d=1.225 g/ml. The concentrated LPs are then applied to a gel filtration column (Bio-Gel A5m) and are eluted in order of decreasing size, i.e. VLDL   LDL   HDL.

This procedure avoids the extensive centrifugation time required by sequential flotation but has the disadvantage that it has not been sufficiently refined for the isolation of the subclasses of the LPs.

IV. ISOLATION OF APOLIPOPROTEINS FROM HDL

ApoA-I and apoA-II are the major apolipoproteins of HDL and are usually derived from isolated HDL. After thoroughly dialyzing HDL against 0.1 M NaCl, 0.05 M carbonate pH 8.6 0.01% azide and 0.01% EDTA, the sample is lyophilized. The dried HDL (5g portions) are transferred to a vessels containing

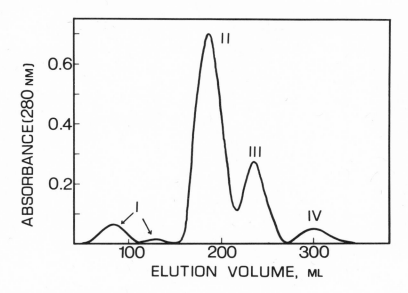

Figure 6: Chromatogram of apoHDL eluted over a 2.5 x 100 cm column of Sephadex G-200 using 8 M urea 0.1 M Tris-HCl pH 7.4 at 10^{-4} EDTA of azide.

about 100 ml of 3:1 ethanol: diethyl ether and shaken for 24 hrs. at 5°. The sample is centrifuged and the ethanol:ether is replaced daily until the protein residue shows the absence of phosphorus (16).After a final extraction with diethyl ether, centrifugation and removal of the solvent, the apoHDL is dried under a gentle stream of nitrogen.

At this point the apoHDL may be solubilized for chromato- graphy in one of several media. These are 6.0 M guanidine HCL, (17) 8.0 M urea (18) or sodium dodecyl sulfate (19). In addi- tion all are run in 0.1 M Tris pH 8.6 containing 0.01% EDTA and sodium azide. Up to 200 mg of apoHDL in 10 ml of the elution media may be applied to a 2.6 x 200 cm column of Sephadex G 150 eluted at a rate of 10-15 ml/hr. A typical chromatogram of apoHDL given in Fig. 6 exhibits four prominent peaks; in order of their appearance they are (I) unidentified material, (II) apoA-I, (III) apoA-II and (IV) apoC. An additional chromatography of the isolated apoA-I on apoA-II over DEAE cellulose in 6 M urea provides a highly purified protein (20). This provides ample amounts of apoA-I and apoA-II for many studies but is not the best source of C-proteins.

V. ISOLATION OF APOLIPOPROTEINS FROM VLDL

Isolation, dialysis and lyophilization of VLDL is the
same as that described for HDL except that the solvent ratio
for delipidation should be ethanol: diethyl ether, 1:3. This
higher proportion of ether is needed to prevent solubilization
of some of the C-proteins (21). After thorough delipidation
and drying the suggested procedure for obtaining the individual C-proteins is the procedure of Brown et al. (22). These
are as follows. (1) dissolve apoVLDL in 0.2 M Tris pH 8.2 containing 0.1 M sodium decyl sulfate. (2) elute the soluble part
of this material through a Sephadex G-150 column with 0.2 M
Tris pH 8.2 and 0.002 M sodium decyl sulfate. The two major
peaks (Fig. 7) are firstly, B-protein and secondly A-protein
and C-proteins. The C-proteins compose 60% of the total VLDL
protein eluted and are freed of detergent by dialysis. The
final step for isolation of the concentrated C-proteins is
chromatography over DEAE in 5.4 molar urea using a salt gradient (22). A typical profile is shown in Figure 8 ; the
first protein to elute from the column is apoC-I which is
followed by apoC-II and finally apoC-III$_1$ and apoC-III$_2$. An
alternative method of isolating C-proteins is isoelectric
focusing (23).

VI. ISOLATION OF APOLIPOPROTEIN FROM LDL

ApoLDL (apoB) has proved to be very difficult to work
with due to its low solubility in the absence of denaturants
such as sodium docecyl sulfate and guanidine:HCl. The void
volume fraction obtained from chromatography of apoVLDL in
sodium decyl sulfate is relatively pure. The best source of
apoB, however, is LDL in which it represents about 25% of the
total lipoproteins weight. This may be isolated by solubilization in deoxycholate or by delipidation with orgainc solvents. Resolubilization is achieved only with great difficulty
even if detergents are used. It is anticipated that the methodology in this are of lipoprotein purification will improve in
the coming years.

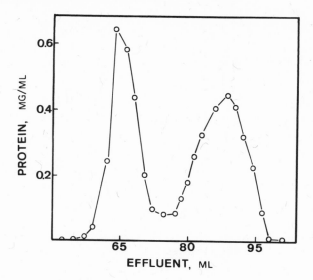

Figure 7: Chromatogram of apoVLDL eluted over Sephadex G-150 in 0.2 M Tris-HCl pH 8.2 containing 0.002 sodium decyl sulfate.

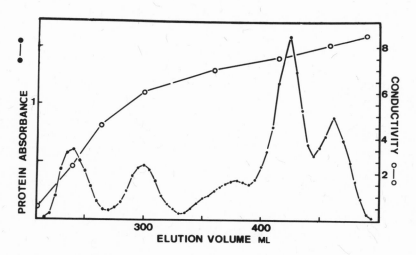

Figure 8: Representative chromatogram of apoC proteins eluted over DEAE in 5.4 molar urea using a salt gradient. The right hand cordinate indicates the salt concentration by conductivity.

VII. APOLIPOPROTEIN ANALYSIS

After pooling and concentrating the individual proteins from HDL and VLDL each pool should be checked for compositional integrity by the following methods:
1. Amino acid analysis consistent with its composition.
2. Urea or sodium dodecyl sulfate-polyacrylamide gel electrophoresis.
3. Reaction with an antibody for a given peptide and the absence of reaction with antibodies prepared from other plasma lipoproteins and albumin.

Acknowledgement

The author is an Established Investigator of the American Heart Association. A portion of this research was supported by a grant from the American Heart Association, the National Institutes of Health (HL-19459) and was developed by the Section on Atherosclerosis, Lipids and Lipoproteins of the National Heart and Blood Vessel Research and Demonstration Center, Baylor College of Medicine, a grant-supported research project of the National Heart, Lung and Blood Institute, National Institutes of Health, Grant No. HL-17269.

References

1. R.L. Jackson, J.D. Morrisett and A.M. Gotto Jr., Physiological Reviews $\underline{56}$ 259 (1976)
2. A.M. Scanu, C. Edelstein and P. Keim in The Plasma Proteins F.W. Putnam, Ed. Academic Press New York (1975) p. 317
3. J.D. Morrisett, R.L. Jackson and A.M. Gotto Jr., Ann. Rev. Biochem. $\underline{44}$ 183 (1975)
4. V.P. Skipski, in Blood Lipids and Lipoproteins: Quantitation, Composition and Metabolism. G.J. Nelson Ed. Wiley New York (1972) p. 471
5. S.J. Friedberg and J.A. Reynolds, J. Biol. Chem. $\underline{251}$ 4005 (1976)
6. H.B. Brewer Jr., S.E. Lux, R. Ronan and K.M. John, Proc. Nat. Acad. Sci. USA, $\underline{69}$ 1304 (1972)
7. H.N. Baker, A.M. Gotto Jr. and R.L. Jackson, Proc. Nat. Acad. Sci. USA, $\underline{71}$ 3631 (1974)
8. M.O. Dayhoff, Atlas of Protein Sequence and Structure $\underline{5}$ Nat. Biomed. Res. Found. Washington, D.C. (1972)

9. J.D. Morrisett, H.J. Pownall, R.L. Jackson, R. Segura, A.M. Gotto Jr. and O.D. Taunton in: Chemistry and Biochemistry of Polyunsaturated Fatty Acids (W.H. Kunau and R.T. Holman, eds.) 1976

10. R.S. Shulman, P.N. Herbert, K. Wehrly and D.S. Fredrickson, J. Biol. Chem. 250 12 (1975)

11. R.L. Jackson, J.T. Sparrow, H.N. Baker, J.D. Morrisett, O.D. Taunton and A.M. Gotto Jr., J. Biol. Chem. 249 5308 (1974)

12. R.L. Jackson, H.N. Baker, E.B. Gilliam and A.M. Gotto Jr., (1977) Proc. Nat. Acad. Sci. USA, in press

13. H.B. Brewer, R. Shulman, P. Herbert, R. Ronan and K. Wehrly, J. Biol. Chem. 249 4975 (1974)

14. A.M. Scanu, J. Lipid Res. 7 285 (1966)

15. L.L. Rudel, J.A. Lee, M.D. Morris and J.M. Felts, Biochem. J. 137 89 (1974)

16. G.R. Bartlett, J. Biol. Chem. 234 466 (1959)

17. J.A. Reynolds and R.H. Simon, J. Biol. Chem. 249 3937 (1974)

18. A. Scanu, J. Toth, C. Edelstein, S. Koga and E. Stiller, Biochemistry 8 3309 (1969)

19. W.L. Stone and J.A. Reynolds, J. Biol. Chem. 250 3584 (1975)

20. B. Shore and V. Shore, Biochemistry 8 4510 (1969)

21. A.M. Scanu and C. Edelstein, Anal. Biochem. 44 176 (1971)

22. W.V. Brown, T.I. Levy and D.S. Fredrickson, J. Biol. Chem. 244 5687 (1969)

23. J.J. Albers and A.M. Scanu, Biochim. Biophys. Acta 236 29 (1971)

THE PLASMA LIPOPROTEINS IN NONHUMAN PRIMATES

V.BLATON

Simon Stevin Instituut
and
Department of Clinical Chemistry, St.Janshospitaal
B-8000 Brugge, Belgium

INTRODUCTION

The need for satisfactory animal models arises in part from
the defects of man as a model for the study of atherosclerosis.
Atherosclerosic lesions in humans develop irregularly over a period
of years. This makes the study of either progression or regression
of the disease in humans very difficult. Animal models provide the
advantage that the lesions of fairly definite age can be induced in
a short period of time so that factors either aggravating them or
causing their regression can be studied with reasonable accuracy.
The search for a completely satisfactory animal model is not finished, but an investigator now has available information concerning
a large number of models from which he can choose. While no animal
model exists that duplicates perfectly the human disease and all of
its effects such as myocardial or cerebral infarcts, an investigator
can choose from a variety of species best suited for studies of the
particular facets of the disease he is interested in.
Because of their close phylogenetic relationship to man, non-
human primates have become of considerable interest in a search for
animal models in which metabolic and pathologic processses thought
to be related to atherosclerosis simulating the situation in man.
Non-human primates are a natural choice since spontaneous atherosclerotic lesions have been described in all species so far studied.
Of approximately 200 species of non-human primates, 12 have been
examined for the prevalence of naturally occuring lesions and in
seven of these species diet induced atherosclerosis has been studied. Although numerous genera of monkeys have not been completely
evaluated, those appearing to have the most potential usefulness as
biomedical models in studying human atherosclerosis research are
the New World Genera Saimiri, Cebus and Pan Troglodytes and the Old

World Genera Papio and Macacus. During the past decade, primary
emphasis has been on Rhesus monkeys (Macaca Mulatta), Squirrel monkeys (Saimiri Sciureus), and baboons (Papio cynocephalus).

The search for the completely satisfactory animal model is not
finished, but in most studies there are not enough detailed biochemical data available to distinguish one species from another. However it is opportune to review the situation at the present time
particularly because we have arrived at a critical point in the future use of these animals in research. This is because of a number
of restraints has been imposed by various governments on the sale
and exportation of non-human primates for research. These apply especially to the rhesus monkey and the baboon. There is a real need
for a summary of the results of experimental work to date and for a
detailed appraisal of the experimental work that might be contempla-
ted for the future. Perhaps the most difficult problem of all is
to decide whether a monkey is essential for the proposed work or
whether some more convenient animal such as the rat would be equally
satisfactory. An important aim of our discussion is to try to resolve this problem of selection of a suitable experimental animal
for research into atherogenesis.

ANIMALS AND DIETS

We studied in our laboratory the plasma lipoproteins of the
chimpanzee, the baboon and the rhesus monkey. The animals are kept
at the Zoo of Antwerp.

Then chimpanzees (white faced, Pan Troglodytes) of mixed sexes
(7 males and 3 females) $\pm$ 10 years of age were given a control diet.
Four animals were given a control diet and six animals were fed an
atherogenic diet. These animals were being investigated as part of
a long term study on the effect of high colesterol diets an atherosclerosis in the chimpanzee and had been consuming the diets for
7-8 years. The controlgroup received 10gm basic control diet per
Kg body weight/day, supplemented with two apples, two bananas and
one-half orange. Vegetables (400g/day), one egg half a liter of
skimmed milk a day were also given. The daily caloric intake amoun-
ted to 1750Kcal/day for the control group. The control diet had a
lipid content of 3gm % with 0.15gm % cholesterol and a protein content of 17gm %. Palmitic (22%), oleic (31%) and linoleic (30%)
were the chief fatty acids in the control diet.

Eighteen young baboons (Papio cynocephalus), three females and
15 males, 12-18 months of age, were divided in equal groups and
given either a control or an atherogenic diet. These animals were
also part of a long term study of the effect of hypercholesterolemic diets on atherosclerosis in the baboon and had been consuming
the diets for 1.5-3 years.

The control diet consisted of rat cake supplemented with vita-
min C. Palmitic (27%) oleic (23%) and linoleic (31%) are the chief
fatty acids in the control diet. The total intake of the diet a-
mounted to approximately 25g/Kg body weight/day. Sixteen male rhe-

sus monkkeys (Macaca mulata) were fed the same diet.   (V.Blaton and
H.Peeters, 1976).

BLOOD SAMPLING

It is always a difficult task to collect blood from the chimpanzee and the baboon.   (Mortelmans J., 1965).
All animals were tranquilized with sernylan (Parke Davies) or with
thalamonal (Janssen Pharmaceuticals) and fasting blood into EDTA
(1mg/ml) was always taken from the femoral artery. Plasma was obtained after centrifugation at 3.000g and used with 6hr for lipoprotein isolation. For lipid and fatty acid analysis samples were
kept at 4°C (Blaton et al, 1974).

BIOCHEMICAL DATA OF THE LIPID AND LIPOPROTEIN ANALYSIS

1. Lipid and lipoprotein analysis

In "Blood lipids and lipoproteins : quantitation, composition
and metabolism", edited by Gary Nelson, we described our integrated
approach to plasma lipid and lipoprotein analysis (V.Blaton and
H.Peeters, 1971). Electrochromatography has been used for years
in our laboratory (1965-1972) as a rapid efficient and inexpensive
means of separating lipoproteins into two groups on a paper curtain.
In a second stage, the fractions are cut out and eluted, their lipids are extracted and the polar and non-polar lipids classes are
separated by thin layer chromatography and FA analyzed by GLC.
Since elution of lipids from lipoproteins, separated by ECG with
chloroform-methanol (2:1, V/V) results in the loss of 10% of phospholipids, a new extraction procedure with chloroform-methanol and
acetic acid was developed. The curtain fragments containing the
high and low density lipoprotein fractions from approximately 2ml
of plasma are lyophilized for 15hr and then extracted with 100ml
chloroform-methanol 1:1 (V/V) for 15hr with constant stirring. After a second extraction for 2 hr with the same solvent, the last
traces of phospholipids are extracted with methanol-acetic acid,
99:1 (V/V). Recovery of individual lipids by elution from the paper was 99.2%. All axtracts are combined and evapored under nitrogen at 40°C. Thin layer chromatography was used for the separation
of polar and apolar lipids with the classical procedures. However,
the apprearence of the Iatroscan using chromaroads a large number
of new applications of TLC are available. In the last months the
Institute developed a new quantitative and qualitative TLC method
on Iatroscan for lipid fractionation (Vandamme et al, 1977).
We started years ago on a program of comparative analysis of primate plasma from animals kept at the Zoo of Antwerp under an identical diet and the same hygienic conditions. In the first experiment following experimental animals are considered chimpanzees,
dwarf chimpanzee, male mountain gorilla, lowland gorillas, baboons,
rhesus and other species.

A comparative study of the lipoproteins separated at that time
on paper shows a progressive increase in the alpha to beta relationship. All primates have more alpha lipoprotein than man. The
mountain gorilla has however lower beta Lp than the lowland gorilla
and slightly more alpha lipoprotein than man. In one of the species
there were two fractionated HDL lipoproteins (Blaton et al, 1968;
H.Peeters et al, 1968).

Excepting the baboon all animals show high total lipids. In
general the relative contribution of each lipid class is of the
same order of magnitude with a few exceptions. The cholesteroles-
ters dominate the picture as could be expected from the absolute
figures, but the phospholipids are percentually more important than
in man, although the C/PL ratio is slightly higher than the normal
human ratio.

Low lipid values are obtained for the baboon. Total cholesterol is mainly transported by LDL in the chimpanzee and in the rhesus. In the baboon however, HDL is the chief carrier of cholesterol. As for man and chimpanzee phosphatidylcholine (PC) is 72% and
is the predominant phospholipid component in the baboon and the rhesus, followed by Sphingomyelin (Sphm) which is much higher in the
new world species (10%). The minor phospholipid subclasses as phosphatidylethanolamine (PE), inositol (PI) and phosphatidyl serine
(PS) are identical in the non-human primates and higher in concentration (mg%) than in man. Differences observed in the PL subclasses are related to structural and compositional differences in the
individual apolipoproteins, which contain potential differences in
functional aspects.

We characterized also the lipid composition in alfa and beta-
lipoproteins, which ressemble the human pattern, except for the
distribution of the phospholipid subclasses. In earlier reports no
differences were described for the percentage lipid composition of
the plasma Lp subclasses in primates, with a high ressemblance to
the human values (Blaton et al, 1976).

2. Differentiation of fatty acids in plasma lipoproteins

The nature of the fatty acids in each lipid class is of particular interest, since the physical and chemical properties of the
lipid molecules depend on the lenght of the hydrocarbons chains and
the number of double bounds. Thus the metabolic properties and
physiological characteristics of lipoproteins may be correlated
with variations in fatty acid composition. The procedure is described by Blaton et al (1974). The fatty acid compositions of serum
lipoproteins from fasting normal human subjects and control primates have been compared. The values obtained over an interval of
several months indicate the constancy during this period. In comparison with chimpanzees and baboons, human subjects demonstrated
a greater content of unsaturated fatty acids in both α-and β-lipo-
proteins.

3. Comparative properties of the major HDL apoproteins in non-
   human primates

The objective of the study on the comparative properties of
the major HDL apoproteins in non-human primates was the isolation
and partial characterization of the plasma HDL lipoproteins in order to elucidate the relationship between the apolipoproteins from
man and non-human primates and to stress the usefulness of the animals in the study of metabolism, structure and function of lipoproteins.

On a modified agarose electrophoresis system, the plasma lipoproteins are separated. Baboon plasma contained VLDL and compared
to the human pattern decreased low density lipoproteins. HDL are
the main lipoproteins in the chimpanzee and the baboon. As a result
the % concentration of the baboon total HDL (74%) and the rhesus
HDL (68%) were much higher than in man (32%) and chimpanzee (53%).
In comparison to man the non-human primates showed an inverse HDL_2/
HDL_3 ratio.

The purified THDL (1.063<d<1.210) from the non-human primates
yielded a percentage protein-lipid distribution similar to the normal data in man and had a protein content of 47.5%. Surprising findings were in the percentage distribution of the phospholipid classes. As for man and as for plasma (PC) is the major PL, but the baboon and the rhesus have higher values. Differences in the non-
human primates were in the higher percentage PC and the lower sphingomyelin (Spm) content for the old world representatives.

The apo-HDL peptides from human and non-human primates were
fractionated on Sephadex and on DEAE-cellulose columns. As we know
that baboon and rhesus are characterized by a high amount of high
density lipoproteins in comparison to man and chimpanzee, it was
tempting to relate the low sphingomyelin to PC-ratio in the former
animals to their lipoproteins.

The major component A-I identical for all species, appears in
different polymorphic forms on DEAE-cellulose using a tris gradient.
ApoA-I accounts for about 60% of apoHDL in man and chimpanzee and
for 70% in the baboon. In all four species considered, apoA-I is
very similar both qualitatively and quantitatively. The second
polypeptide, apoA-II, represents 20% of apoHDL in man and about 12%
in chimpanzee. In addition the chimpanzee apoA-I and apoA-II reac-
ted with monospecific antisera raised in the rabbit against either
of the two human products. In the baboon and the rhesus the apoA-II
is eluted at a different volume from the DEAE-column which means
that it represents some difference with the two other species. Actually the main difference between man and baboon apoA-II appears
to be that the human peptide is present in the dimeric form whereas in the baboon it is a monomer. The similarity between chimpanzee apoA-I and man also extended to the microhetogeneity of apoA-I
which was documented by PAGE-isoelectrofocusing.

The amino acid composition of apoA-I in the four species is
identical (no cysteine, no Il except glutamine and methionine) and

shows a complete resemblence to the major HDL apoprotein between both species.

The amino acid composition of apoA-II in man and chimpanzee is identical. The difference between the dimeric apoA-II in man and chimpanzee and the monomeric apoA-II in the baboon and the rhesus is the presence of arginine and the absence of halfcystine in the baboon and the rhesus apoA-II which is substituted by serine suggesting that the disulfide bridge is not required for the lipid binding functions of the protein.

The apoproteins A-I of the four species were treated with carboxy-peptidase A indicated glutamine, the amino acid most rapidly released as the COOH-terminal residue. Threonine and serine were also rapidly liberated but complete release after 30 minutes of digestion was not achieved. Leucine was released at a slower rate. The kinetics of release of the amino acids suggest a COOH-terminal sequence of LEU-Ser-Thr-Gln for a man and the non-human primates. For the NH_2 terminal sequence analysis, the apoA-I proteins were subjected to 30 cycles of the automatic Edman degradation. Human and chimpanzee have the same sequence. The baboon and the rhesus have identical sequences with the exception of residues 15 and 21. In the human apoA-I there is alanine at residue 15 and valine at residue 21.

The amino acid sequence of HDL apoA-II in man is compared to that of baboon and rhesus. The chimpanzee shows a AA sequence identical to man, while the baboon and rhesus have a 77 residue polypeptide which exhibits a great degree of homology with the monomeric form of human A-II. The major difference between the dimeric apoA-II and the monomeric form is the presence of Arg and the absence of half cystine, which is substituted by serine. Additional human to monkey substitutions are Lys 3 → Glu 3, Seri 40 → Ala 40, Ile 53 → Val 53, Gly 71 → Asp 71, most of which are conservative except for position 3 where glutamic acid replaces lysine. The substitutions occuring in the rhesus and baboon were found to have a higher helical potential than the corresponding residues in the human form. This is especially significant for the alanine predicted helical regions, consisting of residues 25-30 and 33-49 appear to be identical in both the human and rhesus proteins.

The structure of native baboon apoA-I is highly ordered with a helical content of 67% compared to 55% in human apoA-I, which is identical for the rhesus monkey.

ApoA-I at a concentration of 36 M is a monomeric form, above 36 M two distinct components with $S°_{20W}$=3.9, are observed which show an identical AA composition when separated by gel permeation chromatography, suggesting self association of apoA-I. The ultracentrifugation behaviour of rhesus A-I differs markedly from that reported for human apoA-I, where apoA-I self associates in solution according to a monomer dimer tetramer-octamer model, at significantly lower protein concentration than those observed for rhesus apoA-I. In our laboratory, studies on the baboon apoA-I show rather monomer dimer self association as in the human and the affinity for phospho-

lipids is higher for the dimeric apoA-I. The marked difference in
the mode of self association between human and rhesus apoA-I is also reflected by the capacity of each of these apoproteins to bind
lipids in vitro. Studies from Scanu's laboratory have shown that
under comparable experimental conditions of protein and lipid concentrations, rhesus apoA-I binds significantly more lipids than
does human apoA-I, this observation has been related to the degree
of self association of these apoproteins in vitro. It is thus apparent that, as we have pointed out before, the solution properties
of apoA-I and other apolipoproteins must be known before they can be
used for meaningful ligand-binding studies.

        In order to investigate whether monomeric baboon apoA-II pre-
sented the tendency to aggregate reported for human apoA-II, the
concentration dependence of the mean residue ellipticity was followed over a 100-fold concentration range. The negative mean residue
ellipticity at 222nm of monomeric baboon apoA-II is concentration
dependent and increases when the monomer to dimer equilibrium is
shifted towards higher molecular weight species by increasing the
concentration. The maximum value (θ222 = -10.240) agrees with that
measured for the human apoprotein and corresponds to an α-helical
content of 37%. Upon titration with increasing quantities of DMPC
vesicles the mean residue at a ratio of 50mol of DMPC/mol of
apoA-II. The maximal value corresponds to an α-helical content of
54%. This figure higher than that reported for the titration of
human apoA-II with egg PC, suggesting that the structure of baboon
apoA-II is stabilized by phospholipid to a larger extent than that
of the human apoprotein. The cysteine residue in human apoA-II
seems not to be involved in self association nor in phospholipid
binding as its substitution with serine in the baboon does not significantly affect the behaviour of the apoprotein.

        The enthalpy changes which evolved on binding DMPC liposomes
to human and baboon apoA-I are plotted as a function of the phos-
pholipid/protein molar ratio. DMPC binding to baboon apoA-I is an
exothermal process at any phospholipid to protein ratio. The maximal enthalpy change is slightly lower for the baboon than for the
human apoprotein while the complex composition is very similar for
both species. The enthalpy changes which evolved on binding DMPC
to baboon apoA-I are exothermal at low lipid/protein ratios, in contrast with the behaviour of human apoA-I. The enthalpy titration
as resulting human apoA-I DMPC was interpreted as resulting from
the superposition of an endothermal process predominant at high
PL/apoprotein molar ratios. The sigmoidal shape of the curve for
baboon apoA-I indicates that these two effects are also present but
that the magnitude of the endothermal process is less than for the
human apoprotein. This endothermal process was attributed to a lipid induced disaggregation of apoA-I in agreement with other reports.

        The enthalpy changes which evolved on binding DMPC liposomes
to native baboon monomeric and human dimeric apoA-II are studied.
The complex formed between DMPC and either human or baboon apoA-II
has a maximal composition of 35 mol of DPMC/mol of human monomeric

apoA-II and 24 mol for the baboon which are quite similar.
    The heat changes observed on binding shingomyelin liposomes at
41° to human and baboon apoA-I are very low 20Kcal/mole at PL/prot
ratio of 100M/mole. This might result from the small amount of
apoprotein incorporated in the complex which was observed by gradient ultracentrifugation. For apoA-II the enthalpy changes compa-
re with those obtained with DPMC. The results of these experiments
indicate that the PL binding to human and baboon apoproteins is
very similar. The complex composition as well as the enthalpy changes involved in complex formation are of the same order of magnitude. However, the hypothesis remain open for the low SPM content
in the baboon (Blaton et al, 1974, 1977; Rosseneu et al, 1977).

CONCLUSION

    Living primates, a group of mammels which constitute a series
of successively more advanced organisms have proved to be very useful in the study of structure and functions of the plasma lipoproteins. The studies however, reported thus far on the serum lipoprotein distribution of various non-human primates are not sufficient to permit any definite conclusion. Only recently attention
has been directed to the study of the serum lipoprotein apoproteins.
The most detailed studies up to now have been conducted on the HDL
apoproteins of Macacus rhesus (Edelstein et al, 1972; Scanu et al,
1972). Data of baboon and chimpanzee HDL apoproteins were described and for detailed results we refer to the described references.
These lipoprotein particles were found to have properties similar
to those of man.

REFERENCES

  1. V.BLATON, H.PEETERS
     The non-human primates as models for studying atherosclerosis.
     Studies on the chimpanzee, the baboon and the rhesus maccacus.
     Atherosclerosis Drug Discovery Book, 67, 33 (1976).

  2. J.MORTELMANS
     Primates Med., 2, 113 (1969).

  3. V.BLATON, D.VANDAMME, M.VASTESAEGER, J.MORTELMANS, H.PEETERS
     Dietary induced hyperbetalipoproteinemia in chimpanzees :
     Comparison to the human hyperlipoproteinemia,
     Experimental and Molecular Pathology, 20, 132 (1974).

  4. V.BLATON, H.PEETERS
     Integrated approach to plasma lipid and lipoprotein analysis,
     Chapter 6 : in "Blood Lipids and Lipoproteins", G.Nelson, ed.,
     369-431 (1971) J.Wiley and Sons, New York.

  5. D.VANDAMME, V.BLATON, H.PEETERS
     Screening of plasma lipids by thin-layer chromatography with
     flame ionization detection on chromarods,
     Journal of Chromatography, 145 (1978).

6. H.PEETERS, V.BLATON
   Lipid fatty acid relationship in electrochromatographic lipo-
   protein fractions,
   Prog. Biochem. Pharmacol., 4, 144-152 (1968), Karger Basel.

7. V.BLATON, H.PEETERS, G.A.GRESHAM, A.N.HOWARD
   Differential fatty acid composition of alpha and beta lipopro-
   teins in baboons
   Prog. Biochem. Pharmacol., 4, 122-125 (1968), Karger Basel.

8. V.BLATON, R.VERCAEMST, N.VANDECASTEELE, H.CASTER, H.PEETERS
   Isolation and partial characterisation of chimpanzee plasma
   hich density lipoproteins and their apolipoproteins,
   Biochemistry, 13, 1127 (1974).

9. V.BLATON, R.VERCAEMST, M.ROSSENEU, J.MORTELMANS, R.L.JACKSON,
   A.M.GOTTO Jr., H.PEETERS
   Characterization of baboon plasma high density lipoproteins
   and of their major apoproteins,
   Biochemistry, 16, 2157 (1977).

10. M.ROSSENEU, V.BLATON, R.VERCAEMST, F.SOETEWEY, H.PEETERS
    Phospholipid binding and self-association of the major apopro-
    tein of human and baboon high density lipoproteins,
    Eur. J. Biochem., 74, 83-88 (1977).

ACKNOWLEDGEMENT

     The author wishes to thank Dr.H.Peeters, Director of the
Simon Stevin Instituut and Dr.Sc.M.Rosseneu for the helpful discussion.
     This research was supported by grants of NFGWO (Brussels,
Belgium, grant nr 1026 and nr 30.0175).
     I am grateful to Mrs.N.Vinaimont, Mr.D.Vandamme, Mr.B.Declercq
and Mr.R.Vercaemst for their excellent technical assistance.

Structure of the Lipoprotein Molecule

PHYSICAL PROPERTIES, CHEMICAL COMPOSITION AND STRUCTURE

OF CIRCULATING LIPOPROTEINS

Ferenc J. Kézdy

Department of Biochemistry
University of Chicago
Chicago, Illinois 60637

Before discussing the structure of lipoproteins, a brief survey of the surface properties of their major constituents is in order. This will be followed by a discussion of the basic thermodynamic laws governing small spherical particles. The main discussion will be concerned with the gross structural features of lipoproteins, which can be predicted from the regularities observed in chemical composition as a function of particle size.

I. SURFACE PROPERTIES OF PHOSPHOLIPIDS AND APOPROTEINS

A. Phospholipids

The pronounced amphiphyllic character of phopholipids with moderate or long aliphatic chains (lecithins, sphingomyelins, etc., with more than eight carbons in the side chains), allows them to form unusually stable mono-molecular layers at the air-water interface (1). The molecular packing and the pressure dependency of the compressibility of these monolayers depends on the degree of unsaturation of the aliphatic chains. Saturated chains, such as the ones in palmitic or stearic acid, are conducive to tightly packed, quasi-solid monolayers, where the molecules are aligned on the surface with their side chains in a hexagonal packing. The presence of double bonds in the side chain, especially the <u>cis</u> bonds of oleic and linoleic acid, results in less well packed layers, where the fliudity of the monolayer is maintained even at very high surface pressures. Typically, unsaturated phospholipids occupy a limiting area of the order of 65-70 $\overset{o}{A}$/molecule under surface pressures of 30-40 dynes/cm. Even under these conditions, the monolayer is liquid, i.e. the chain-

chain interactions are limited to random hydrophobic forces similar to the ones observed in liquid hydrocarbons.

Surface potential measurements also indicate the absence of well-defined phase transitions over a wide limit of surface pressures for unsaturated phospholipid monolayers. It is this property of being able to form layers which remain liquid over a large range of conditions, which renders unsaturated phospholipids eminently suitable as basic constituents of biological surfaces. Liquid phospholipid monolayers are readily miscible with other monolayer-forming molecules, such as cholesterol and proteins. At low surface pressures (2 to 10 dynes/cm) the addition of cholesterol exerts a "condensing effect" i.e. the average area occupied in the mixture by each molecule is decreased by some 10 to 20% (2). The condensing effect is much less pronounced at high surface pressures and the molecules tend to occupy their respective limiting areas, about 40 Å^2/molecule for cholesterol.

B. Apolipoproteins

At the air-water interface, the surface properties of apolipoproteins are rather unique in comparison with other proteins. Globular proteins, such as chymotrypsin lysozyme, or subtilisin, form quasi-irreversibly denatured monolayers with low collapse pressures (of the order of 5 dynes/cm). The formation of these monolayers on the surface of the protein solution is slow and presumably the denaturation of the protein is rate limiting. In contrast, apolipoproteins apo-A-I and apo-A-II form monolayers with a purely diffusion-controlled rate from dilute solutions (10^{-7} to 10^{-8}M), the formation of the monolayer is reversible, and the limiting pressures are of the order of 20 to 25 dynes/cm (3). The partitioning of the protein between the solution and the surface obeys a thermodynamic equilibrium. For each concentration in solution corresponds a given surface concentration, and the equilibrium can be obtained either by starting with the protein in solution only or in the surface monolayer only. At surface pressures of less than 2 dynes/cm the reversibly formed protein monolayer is unstable and it slowly denatures, until the protein reaches very large limiting areas. At higher surface pressures, apo-A-I and apo-A-II occupy areas corresponding to 15 Å^2/residue. Finally, the apoproteins are able to reach the air-water interface even when the surface is covered by a phospholipid monolayer of moderate pressure. The penetration of the protein into the phospholipid layer is indicated by a rapid increase in surface pressure, until a limiting pressure of 20 to 25 dynes/cm is reached (3,4).

II. THERMODYNAMIC RELATIONSHIPS   (5)

Circulating lipoproteins are spherical particles with diameters ranging from 10^{-7} to 10^{-8} cm. The small size and the large curvature of the surface result in far-reaching modifications of the thermodynamic properties, as compared with large particles of the same chemical composition. For a liquid phase containing a solute Gibbs-Duhem equation states that

$$-\Gamma = (C/RT)(\partial\gamma/\partial C)$$

where Γ is the surface excess of the solute in moles/cm^2, γ is the surface tension of the solution in dynes/cm and C is the molarity of the solute. Thus the concentration of the solute at the surface is different from that in the bulk phase and this difference is reflected in the concentration dependency of the surface tension of the solution. Experimental observation of the latter indicates that for aqueous solutions the surface excess is negative for salts, since the surface tension increases with increasing salt concentrations. For organic molecules in general the surface excess is positive; their local concentration close to the surface is larger than in the bulk phase. By measurment of the optical anisotropy of the solutions in the surface region it is possible to show that the thickness of the "Gibbs layer" is somewhere between 20 and 100 A, depending somewhat on the solute. Thus, the anisotropy due to the surface is transmitted to molecules several molecular diameters away from the surface and these asymmetric forces profoundly influence the chemical and physical properties of all molecules close to the surface. Since for lipoproteins the majority of the components are located in the Gibbs layer, the extrapolation of properties observed in bulk solutions to lipoproteins must be accomplished with due caution.

Lipoproteins are spherical particles indicating that a positive interfacial tension between the water and the particle decreases the surface until a minimal surface corresponding to an energy minimum is reached. At equilibrium, the compressive forces of the surface tension are compensated by an internal pressure. This pressure inside the particle can be calculated from the Young-Laplace equation:

$$\Delta P = 2\gamma/r$$

where ΔP is the difference of pressures inside the particle and in the solution and r is the radius of the spherical particle. Application of this equation to lipoproteins shows that if the interfacial pressure is 20 dynes/cm for HDL the internal pressure is of the order of 100 atm.

A further consequence of the surface curvature is the modification of the chemical potential of the molecules in a small sphere. The Kelvin equation states, that

$$\mu-\mu_f = 2\gamma V/r$$

where μ is the chemical potential of the liquid in a spherical particle, μ_f is that of the same compound in a large bulk phase, and V
is the molar volume of the compound. Thus, a decrease in particle
size results in an increase of the chemical potential, i.e. the
partial molar free energy. Since the chemical potential governs all
chemical equilibria and many of the physical properties, the chemical
reactivity and the transferability of a molecule in a small particle
is much larger than in the bulk phase. As an extension of the above
relationship, the equation derived by V.K. La Mer and R. Gruen (6)
states that

$$\ln X_r = \ln X_f - (2\gamma V/RT)(1/r)$$

where X_r is the mole fraction of a solute in a spherical particle of
radius r and X_f is that of the same solute in the bulk phase (6).
This equation shows that when a solute is distributed between particles of different size, the concentration of the solute decreases
with decreasing radii.

In summary, the large curvature and the small size of lipoproteins confers special properties to lipoprotein constituants, properties not directly measurable using the same components in a bulk
phase. Most importantly, the thermodynamic relationships point to
the important role of the lipoprotein surface as a determinant of
the physico-chemical properties of the constituant molecules.

III. CHEMICAL COMPOSITION AND STRUCTURE OF LIPOPROTEINS

The chemical composition of circulating lipoproteins show a
considerable variation in the percentage of the major components

Table I.   CHEMICAL COMPOSITION BY DRY WEIGHT PERCENTS
                OF HUMAN SERUM LIPOPROTEINS

Particle	$MWX10^{-6}$	Protein	Phospho-lipids	Trigly-cerides	Cholesteryl ester	Free cholesterol
$VHDL_1$	0.15	62	28	4.6	3	0.3
HDL_3	0.18	55	23	4.1	12	2.9
HDL_2	0.36	41	30	4.5	16	5.4
LDL	2.3	21	22	11	37	8
VLDL	19.6	8	18	50	12	7
CHYL	504	2	7	84	5	2

for the different density classes of lipoproteins. As shown in
Table I, (7) increasing particle size is accompanied by a monotonous
decrease in protein content, an increase in triglyceride, and a decrease in phospholipids for the larger particles. On the other hand,
the cholesteryl ester and free cholesterol contents pass through a
maximum for LDL.

These regularities in the composition suggest that there is a
casual relationship between particle size and chemical composition,
i.e. for a given particle size there is just one chemical composition which satisfies the requirements imposed by the structure. If
this is indeed the case, then the structural features can be deduced
from the compositional data. For such an undertaking, the data in
Table I were first converted into numbers of molecules of each constituent per particle, using the estimated molecular weights of 850
for triglycerides, 775 for phospholipids, 650 for cholesteryl esters
and 387 for cholesterol (Table II). Since proteins in the different lipoproteins have widely different molecular weights, the per-
cent of protein was expressed in terms of the number of amino acids
per particle, using an average residue weight of 100. In addition
to the data presented in Tables I and II the numerical analysis also
included some twenty other lipoprotein compositions reported for
subclasses of VLDL and lipoproteins of a variety of vertebrates (8).

The starting point of the structural analysis was the determination of space filling requirements of the hydrophobic constituents
i.e. triglycerides and cholesteryl esters. If these constituents
occupy a hyrophobic spherical core at the center of the lipoprotein,
then the sum of their volumes should be linearly related to the cube
of the radius of the particle (r) decreased by the thickness of an
amphiphillic surface layer (x). Thus the number triglyceride mole-

Table II.   MOLECULES OF COMPONENTS PER PARTICLE

Particle	Amino acid	Phospho- lipid	Free cholesterol	Cholesteryl ester	Trigly- ceride	Equivalent radius (A)
$VHDL_1$	961	56	1.2	8	8.3	37.5
HDL_3	963	51	13	32	9.5	39.2
HDL_2	1476	137	50	90	19	50.8
LDL	4830	653	475	1310	298	95.9
VLDL	15656	4545	3539	3600	11500	200
CHYL	102000	45160	25840	27700	507000	600

cules per particle (n_{tg}) and that of cholesterol esters (n_c) should obey the equation:

$$n_{tg}v_{tg} + n_{ce}v_{ce} = (4/3)\pi(r-x)^3 \tag{1}$$

where v_{tg} and v_{ce} are the molecular volumes of triglycerides and cholesteryl esters respectively. Since the ratio v_{ce}/v_{tg} should be equal to the ratio of the molar volumes of the two species of molecules (= 0.68) Equ. 1 can be transformed into the linear form:

$$(n_{tg} + 0.68\, n_{ce})^{1/3} = (4\pi/3v_{tg})^{1/3}(r-x). \tag{2}$$

The data for all lipoproteins indeed yielded an excellent straight line when analyzed according to Equ. 2 (correlation coefficient = .997) and the following parameters were obtained: v_{tg}=1556 Å^3/ molecule; v_{ce}=1068 Å^3/molecule and x=20.5 Å. Thus for all lipoproteins, the triglycerides and cholesteryl esters occupy a spherical core of diameter r−20.5 Å and the packing of these hydrophobic molecules corresponds to that of the pure constituents in the liquid state.

Preliminary analysis indicated that the surface of the hydrophobic core is too small to accomodate the proteins as well as the amphiphillic phospholipids and cholesterol. On the other hand the surface of the core is correlated with the number of phospholipid (n_{pl}) and cholesterol molecules (n_c) alone. The following equation was tested:

$$s_{pl}n_{pl} + s_c n_c = 4\pi(r-x)^2 \tag{3}$$

where s_{pl} and s_c are the molecular areas of phospholipids and cholesterol respectively. Taking for the ratio s_c/s_{pl} the value observed for their ratio in a tightly packed monolayer at the air-water interface (=40.5 Å /71 Å =o.57), Equ. 3 can be transformed into a linear relationship:

$$(n_{pl} + 0.57n_c)^{1/2} = (4\pi/s_{pl})^{1/2}(r-x). \tag{4}$$

Numerical analysis again indicated that Equ. 4 is an excellent description of the phospholipid and cholesterol composition of all lipoproteins, with the following numerical parameters: s_{pl}=68.5 Å^2/molecule; s_c=39.1 Å^2/molecule and x=20.2 Å. Thus only phospholipid and cholesterol are in contact with the hydrophobic core, and these surface molecules are packed on the inner surface to the same extent as that found in their highly compressed monolayers at the air-water interface.

In the light of the preceding considerations, the outer surface of the lipoprotein particle in contact with water should be occupied

then by proteins as well as by phospholipids and cholesterol. Preliminary analysis indicated however, that the area occupied by cholesterol on the outer surface is negligibly small, i.e. free cholesterol is unable to reach the lipoprotein-water interface. Under
these conditions the surface-filling requirements impose the following relationship:

$$s'_{aa}n_{aa} + s'_{pl}n_{pl} = 4\pi r^2 \tag{5}$$

where s'_{aa} and s'_{pl} are the area occupied by an average amino acid
residue and phospholipid molecule respectively.

Equ. 5 can be transformed into the linear form:

$$r^2/n_{aa} = s'_{aa}/4\pi + (s'_{pl}/4\pi)(n_{pl}/n_{aa}). \tag{6}$$

The experimental compositional data indicated excellent correlation
with Eq. 6 and the following parameters were obtained: s'_{aa}=15.6 $\overset{o}{A}{}^2$/
residue and s'_{pl}=62.7 $\overset{o}{A}{}^2$/molecule. Thus, the lipoprotein-water interface is occupied exclusively by phospholipids and proteins, and cholesterol must be covered by polypeptide chains.

Finally, the cholesterol content of all lipoproteins is in excellent agreement with the La Mer-Gruen equation. Indeed, numerical
analysis of the cholesterol content of the surface layer of lipoproteins indicates that cholesterol is a true solute in this layer
and that the interfacial tension between water and lipoproteins is
39 dynes/cm. It should be pointed out that the data for chylomicrons deviate from the theoretical equation, suggesting that the rapid degradation of chylomicrons prevents their equilibration with
the other lipoproteins in the circulating plasma.

The preceding analysis demonstrates, that the chemical composition of all lipoproteins is in excellent agreement with the space-
and surface-filling requirements of a structural model composed of
a hydrophobic core and amphiphillic surface monolayer. There never
was much doubt about the triglycerides and cholesteryl esters occupying a centrally located core, but the present model stresses their
clear segregation from the rest of the lipoprotein components. Also,
the analysis reveals the existence of a well defined surface monolayer, the components of which do not interpenetrate with triglycerides and cholesteryl esters. The numerical results obtained for
the thickness of this monolayer (20 $\overset{o}{A}$) and the various molecular
area are in excellent agreement with those observed for monolayers
of the same components at the air-water interface. These eminently
reasonable numerical results further support the validity of the
proposed model.

The essential features of the proposed model are in agreement
with the large majority of experimental observations concerning

the physical and chemical properties of lipoproteins. On the other hand, a space- and surface-filling analysis is inherently unable to make predictions concerning the chemical and physiological role of the variety of subclasses of components, such as the sphingomyelins, lecithins and phosphatidyl ethanolamines, or the different apoproteins. It is most likely, however, that the structural role of these subclasses is very similar, if not identical.

ACKNOWLEDGEMENTS

The work presented in this article was generously supported by Grants 1 P17 HL 15062 and 1 P01 HL 18577 from the National Heart and Lung Institute, U.S. Public Health Service. The author wishes to thank Dr. Betty W. Shen and Dr. Angelo M. Scanu for their contributions.

REFERENCES

1.)   For a review, see e.g. G.L. Gaines, Jr.: "Insoluble Monolayers at Liquid-Gas Interfaces". Interscience, N.Y., 1966.
2.)   D. Chapman, N.F. Owens, M.C. Phillips and D.A. Walker; Biochim. Biophys. Acta 183, 458 (1969).
3.)   B.W. Shen, A.M. Scanu and F.J. Kézdy; Circulation, 7, Suppl. IV 218 (1973).
4.)   M.C. Phillips, M.T.A. Evans and H. Hauser; Adv. in Chemistry Series 144, 217 (1975).
5.)   For a review, see e.g. A.W. Adamson: "Physical Chemistry of Surfaces". 3rd Edition, Interscience, N.Y., 1976.
6.)   V.K. La Mer and R. Gruen; Trans. Farad. Soc. 48, 410 (1952).
7.)   From A.M. Scanu and A.W. Kruski: "International Encyclopedia of Pharmacology and Therepeutics". Pergamon, Oxford Vol. 1, sect. 24, 21 (1975).
8.)   B.W. Shen, A.M. Scanu and F.J. Kézdy; Proc. Natl. Acad. Sci. USA 74, 837 (1977).

THE PHYSICO–CHEMICAL PROPERTIES OF LIPIDS IN LIPOPROTEINS

M.C. Phillips

Unilever Research Laboratory, The Frythe, Welwyn, Herts., U.K.

I. INTRODUCTION

As can be seen from Table 1, serum lipoproteins comprise a variety of lipid molecules. The density of the particles decreases as the content of neutral lipid increases. Thus, low density lipoproteins (LDL) and chylomicrons contain a preponderance of triglycerides and cholesterol esters which are immiscible with water. Apart from chylomicrons, the various lipoprotein classes contain roughly the same proportion of polar, water–dispersible lipids (phospholipids). Clearly, any physico–chemical discussion of the properties of lipoprotein lipids must consider their mixing behaviour, as well as the properties of the pure, anhydrous species and their interactions with water.

Table 1

The Principal Components of Human Serum Lipoproteins[a]

Component[b] (% w/w)	HDL$_3$	LDL	VLDL	Chyl
Protein	~ 56	21	10	2
Cholesterol	3	8	10	3
Cholesterol Esters	12	37	5	2
Glycerides	4	11	~ 55	~ 88
Phosphatidylcholine	17	14	~ 12	3
Sphingomyelin	3	6	~ 5	2
Non–esterified fatty acids	2			

a. Adapted from data in ref. 1,2
b. Components present at ≲ 1% level are not included

91

This paper first considers the physical states which hydrocarbon chains can adopt, dealing in particular with polymorphism
and the effects of temperature. After this, the consequences of
the hydrophobicity of hydrocarbon chains for the phase behaviour
and surface-activity of phospholipids in water is dealt with.
Finally, the miscibility of polar and neutral lipids is discussed;
particular emphasis is paid to the effects of altering the ratio
of phosphatidylcholine to cholesterol on the emulsification of
neutral lipids in water.

II. THE BEHAVIOUR OF HYDROCARBON CHAINS

The C-C single bonds in a free alkane chain can take up
three stable conformations of which the two gauche conformations
are higher in energy than the trans conformation (for reviews,
see ref. 3-5). In a condensed lipid phase, intra- and intermolecular steric restrictions limit the number of possible chain
conformations. At low temperatures the chains are likely to be
in the low energy all-trans conformation (Fig. 1). As the
temperature is raised, there will be some probability of gauche
rotations about C-C bonds with energies typically about 0.5 kcal
mol^{-1} appearing in the chains. In a molecule such as butane,
where the energy barrier preventing rotation between trans and
gauche conformations is ca. 3.6 kcal mol^{-1}, the jump frequency

Fig. 1.   Schematic representations of hydrocarbon chain configurations.
The zig-zag lines represent the chains of methylene groups
which on the left are in the all-trans conformation. The
remaining chains have several gauche rotations in the lower
half and show different angles of tilt.

between the two states is ca. 10^{10} s^{-1}. Steric interactions between chains prevent most single bond trans-gauche changes until relatively high temperatures when a cooperative melting of the chains occurs.

Table 2

Thermodynamic Data for the Thermal Transitions of Compounds Containing n-Hexadecane Chains

Compound	Ordered Form*	Melted Form	T_c °C	ΔH kcal mol^{-1}	ΔS cal mol^{-1} deg^{-1}	Ref.
n-hexadecane	β crystal	isotropic liquid	18	13	44	27
palmitic acid	β "	"	63	13	39	58
tripalmitin	β "	"	66	43	127	
	β′ "	"	45	30	94	58
cholesteryl palmitate	crystal	cholesteric mesophase	78	15	42	31
	cholesteric mesophase	isotropic liquid	80	0.3	1	31
dipalmitoyl phosphatidyl-choline						
– monohydrate	β′ crystal	smectic mesophase	65	5	14	27
– in excess water	β′ gel	"	41	9	28	27

*The nomenclature of Luzzati[14,46] and coworkers is employed here.

Table 2 contains thermodynamic data describing the chain melting of a series of lipids containing n-hexadecane chains. Since all the compounds melt at higher temperatures (T_c) than n-hexadecane it is clear that the presence of polar groups stabilises the crystal phases of these lipids. The T_c increases with increasing chain length and the more unsaturated the hydrocarbon chains the lower is the T_c; cis double bonds have much greater effects than trans double bonds[6]. The predominant fatty acids in serum lipoproteins are palmitic (n-hexadecanoic acid), oleic (cis-9-octadecenoic acid) and linoleic (cis-9-,cis-12-octadecadienoic acid)[2]; the melting points of these species are 63 (β crystal-Table 2), 16 (β crystal) and –5°C, respectively.

Besides the contribution from the increase in internal energy of the hydrocarbon chains due to rotational isomerism, the rise in enthalpy (ΔH) at the T_c contains a contribution from the change in intermolecular cohesive and repulsive forces arising from expansion

of the lipid (the pΔV work term is negligible). It is apparent
from Table 2 that ΔH per hydrocarbon chain for the transition from
β crystal to isotropic liquid (melt) is constant at about 13 kcal
mol^{-1}; thus ΔH is proportional to the number of single C–C bonds
involved in the transition. The sharpness of the phase transition
is of interest because it depends on the number of molecules
forced to cooperate in the transition; the latter number is called
the cooperative unit and for a first order transition is infinity.
However, in real crystals imperfections can cause local melting so
that the cooperative unit is smaller than the total solid. The
Van't Hoff enthalpy (ΔH_{VH}) relates to the reversible, two-state
transition of a cooperative unit. It is possible to derive ΔH_{VH}
from differential scanning calorimetry traces of excess specific
heat through the temperature range of the transition because

$$\Delta H_{VH} \approx 6.9 \frac{T_m^2}{\Delta T_{\frac{1}{2}}} \qquad\qquad [1]$$

where T_m = mid-point temperature and $\Delta T_{\frac{1}{2}}$ = half-height width in
degrees of the specific heat curve[7]. Integration of the area under
the excess specific heat curve gives the calorimetric enthalpy of
transition (ΔH_{cal}); this corresponds to the transition enthalpy
per single molecule. Thus, the ratio of ΔH_{VH} to ΔH_{cal} gives a
convenient measure of the cooperativity of the transition and may
be interpreted as the number of lipid molecules in the average
cooperative unit. Using this approach, the cooperative unit in
synthetic phosphatidylcholine bilayers when dispersed in water as
liposomes is ~ 100 molecules[8]; this is not necessarily the overall
size of a local patch of solid or fluid molecules[9]. Since the
cooperative unit in lipoproteins cannot be larger than a single
particle and these frequently contain less than 100 lipid molecules of a given type[2], any chain-melting transitions induced in
lipoproteins are likely to be of low cooperativity and therefore
broad relative to the transitions in liposomes containing a single
type of phospholipid molecule. This means that there will be a
temperature range in which crystalline and melted lipid molecules
coexist in lipoprotein particles.

The data for tripalmitin in Table 2 show that different
chain-melting transitions can be observed at 45 and 66°C. This
arises because lipids such as tripalmitin can adopt different
crystal forms i.e. they exhibit polymorphism.

<u>Polymorphism</u>

Lipids generally crystallise so that the molecules are
arranged in layers bounded by planes of end-groups (terminal methyl
groups, polar head groups). The molecular arrangement is the
result of the interplay of the packing requirements of the polar

groups and the long hydrocarbon chains.   These effects can lead to
a variety of crystal structures and polymorphism is common with
lipids (for reviews, see ref. 10-13).   When one of two polymorphs
is thermodynamically unstable at all temperatures below T_c, the
system is said to be monotropic.   Only the transformation from
metastable to stable polymorph can be observed directly and the
metastable polymorph can often be obtained only by transformation
from metastable (undercooled) liquid.   When each of two polymorphs
is thermodynamically stable in a definite range of temperature and
pressure they are said to be enantiotropic; a transition between
the two polymorphs occurs at a definite temperature.   It should
be noted that the distinction between monotropic and enantiotropic
polymorphism can often be blurred by kinetic effects.

    The results of X-ray powder studies of lipid crystals have
shown that n-alkane chains adopt the trans configuration and
various angles of tilt with respect to the plane of the end groups,

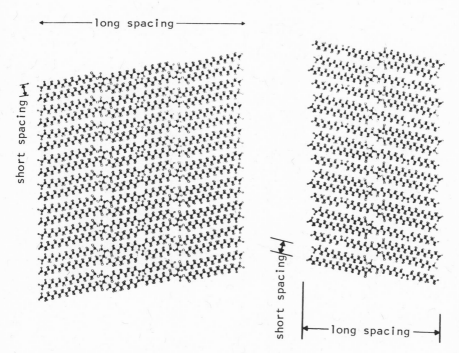

Fig. 2. The **right** diagram illustrates the double chain packing of
        the triglyceride (POS) containing palmitic, oleic and
        stearic acid chains in the 1-,2- and 3-positions,
        respectively.   The left diagram represents triple chain
        packing of POS.   Clearly, the X-ray long spacings are
        quite different in the two cases.

but the conformation of the polar groups cannot be obtained by this technique.   Several nomenclatures have been used to describe the various polymorphs and we shall follow Luzzati[12] in the subsequent brief description of the structures adopted by hydrocarbon chains. In the β form the hydrocarbon chains are stiff and fully extended and organised with rotational disorder according to a two–dimensional hexagonal lattice; the β′ conformation is similar except that the chains are tilted with respect to the normal to the plane of the lamellae (cf. Fig. 1; see ref. 12,14).   The exact crystal packing often depends upon the solvent used for recrystallisation. Examples of some polymorphic forms of triglycerides are depicted in Fig. 2 (cf. 11).   The chain conformation above the melting point is designated α.

   Knowledge of the detailed conformation of the whole lipid molecules in crystals requires the complete elucidation of the structure of a single crystal.   By way of example, we describe briefly the results of an X–ray diffraction analysis by Hitchcock et al.[15,16] of 1,2 dilauroyl–DL–phosphatidylethanolamine crystals grown from glacial acetic acid.   The monoclinic crystals contained one molecule of acetic acid per lipid molecule and were of mediocre quality by ordinary crystallographic standards in that the long range order was poor.   The one–dimensional electron density pro–

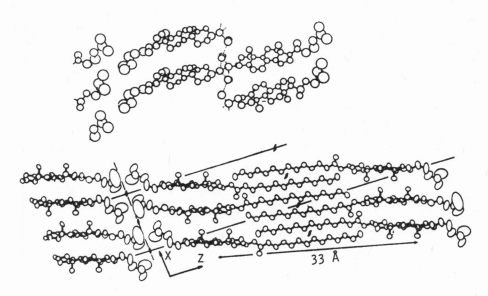

Fig. 3.   Top: crystal structure of cholesterol monohydrate (from
          ref. 19).   Bottom: crystal structure of cholesteryl
          myristate (from ref. 20).   The atoms have a 50%
          probability of being enclosed by the circles or ellipsoids.

files are characteristic of those from phospholipid bilayer
membranes (cf. 17,18). Thus, the intermolecular packing of the
phosphatidylethanolamine molecules produces the classical lipid
bilayer structure, adjacent bilayers being separated by acetic
acid molecules of crystallisation. The lipid chain axes are
essentially parallel to one another and there is a progressive
increase in disorder towards the centre of the bilayer. Within
one section of bilayer, the separation of the centres of gravity
of the polar groups is 39.0Å. The P–N axis of the zwitterion is
approximately parallel to the plane of the bilayer and there is
an intermolecular hydrogen bond of length 2.79Å between the primary
amino group and the phosphate group.

Recent X–ray single crystal studies by Craven and coworkers[19,20]
have yielded the crystal packings of cholesterol monohydrate and
cholesterol myristate (see Fig. 3). The structure of cholesterol
monohydrate shows a stacking of bilayers which are 33.9Å thick.
The ring systems of all cholesterol molecules are nearly parallel
and the C–3 hydroxyl groups and water molecules are hydrogen–bonded
to form a pleated sheet at the bilayer interface. The C–17 chain
atoms have large thermal mean square amplitudes (Fig. 3) and a
reversible endothermic ($\Delta H = 0.7$ kcal mol^{-1}) transition occurs
between 35 and 40°C which is related to a change in the packing of
these terminal CH_3 groups[21]. As shown in Fig. 3, these groups are
also highly mobile in cholesterol myristate crystals. In this case,
the structure consists of stacked bilayers 50.7Å thick in which
cholesterol myristate molecules are packed in an antiparallel
fashion so that the cholesterol residues are adjacent with the
myristate chains interdigitated. The latter chains are all trans
and so in the β configuration.

Since the hydrocarbon chain packing is the same in the β′
forms of tripalmitin and dipalmitoyl phosphatidylcholine, the fact
that in the latter case ΔH and ΔS for chain melting are lower
(Table 2) shows that the melted form of phosphatidylcholine (i.e.
the smectic mesophase) is more ordered than liquid tripalmitin.
The reasons for this are discussed in the next section.

Thermotropic Mesomorphism

Many compounds do not pass directly from the crystal to the
liquid state on heating, but exhibit one or more intermediate
liquid–crystalline or mesomorphic states[22-24]. Three types of
mesomorphic states are generally recognized and these are depicted
in Fig. 4. Smectic liquid crystals are made of sheets of orientat-
ed molecules; fluidity arises because the sheets are flexible and
can slide over one another. Nematic liquid crystals are not as
highly ordered as smectic mesophases; the molecules are arranged
with their long axes parallel but not separated into layers. The
molecules apparently can be drawn past one another in the direction

of their long axes, a degree of freedom which is excluded in smectic
phases by the comparatively strong attractions between the ends of
the molecules. As the name implies, the cholesteric mesophase is
exhibited by many cholesterol derivatives; the molecules in each
layer are parallel but because of the shape of cholesterol molecules, the orientation of the molecules in each layer is displaced
progressively. The overall displacement traces out a helical
path and this has an effect on light transmitted perpendicular to
the molecular layers so that circular dichroism occurs.

Only phospholipids and cholesterol esters of the lipids which
occur in serum lipoproteins (Table 1) form mesophases; both exhibit
smectic mesophases and cholesterol esters also form cholesteric
liquid crystals. The thermotropic mesomorphism of phospholipids
has been investigated in great detail (for reviews, see 3,6,12,13,
18,25). The lamellar phase is comprised of bimolecular leaflets
of hydrocarbon chains covered on both sides by the polar groups.

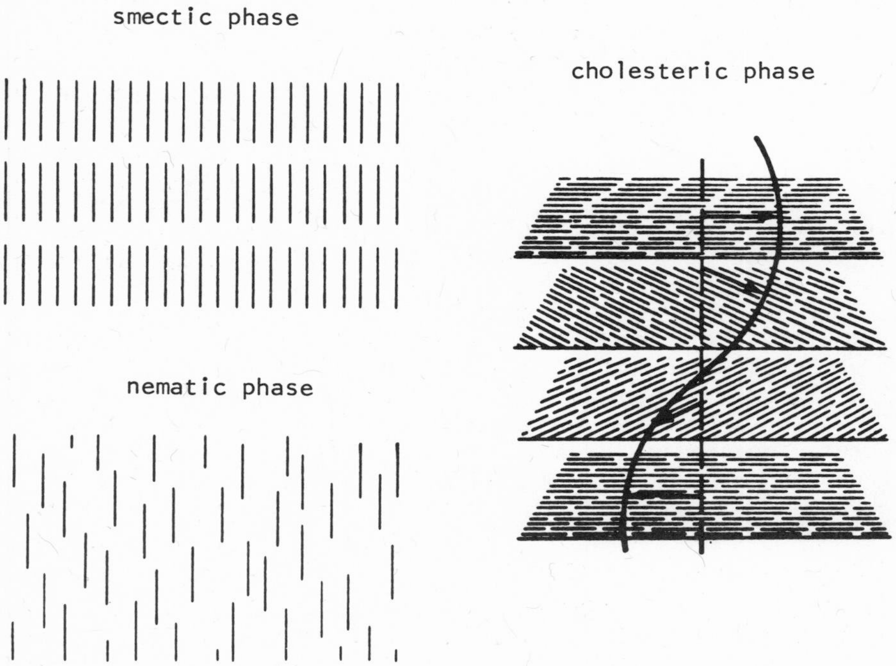

smectic phase

cholesteric phase

nematic phase

Fig. 4.   Schematic representations of the molecular arrangements
           characteristic of the various forms of thermotropic
           liquid crystal. The helicity of the overall displacement
           of molecular orientation from layer to layer in the
           cholesteric mesophase is indicated.

The lamellae are parallel and equidistant. In the hexagonal phase the polar groups form cylinders of indefinite length which are packed in a two-dimensional hexagonal lattice, the hydrocarbon chains occupying the region between the cylinders. In the cubic phase, identical rodlike elements are linked three by three into two three-dimensional networks which are unconnected but mutually interwoven. The physical state of the hydrocarbon chains in the bilayers of the predominant smectic phase is of particular interest. As can be seen from Table 2, ΔH and ΔS per hydrocarbon chain for the transition of dipalmitoyl phosphatidylcholine from the β' crystal to smectic mesophase are approximately half the equivalent figures for chain-melting of tripalmitin where the isotropic liquid is formed. This reduction in the configurational freedom of the hydrocarbon chains in anhydrous phospholipid mesophases, as compared to the fluidity of liquid n-alkanes at the transition temperature, is a consequence of the order imposed by the polar group lattice[3,26,27]. The chains are anchored at one end to the most rigid part of the bilayer, the glycerol backbone of the phosphatidylcholine molecules, and in this region the possibility of trans-gauche isomerisations is reduced because the steric restriction is greatest in this region[28]; the configurational freedom and mobility of the chains increases towards the centre of the bilayer so that the terminal methyl groups have correlation times of $\sim 10^{-10}$ s which is similar to that in liquid n-alkanes[29].

Cholesterol does not form any mesophases and melts directly to an isotropic liquid at 147°C. In accord with the observation[30] that ΔH and ΔS for the melting of rigid molecules is generally lower than for flexible molecules because of the lower possible configurational disorder, the ΔH of 6.5 kcal mol^{-1} and ΔS of 15 cal mol^{-1} deg^{-1} for melting cholesterol[31] are lower than the equivalent figures for lipids such as n-hexadecane or palmitic acid (Table 2). Cholesterol esters behave differently in that they form mesophases (for reviews, see 23,31,32); some of the thermodynamic data for cholesteryl palmitate are presented in Table 2. A cholesteric mesophase is formed on heating the crystals to 78°C and isotropic liquid forms at 80°C. Cooling the isotropic liquid leads to the reversible formation of cholesteric mesophase at 78°C which converts to smectic mesophase at about 74°C (ref. 31); this phase is monotropic and gradually converts to the crystal phase. As with the melting of simple fatty acids, the transition temperatures of cholesteryl esters are a function of the degree of unsaturation of the fatty acid moieties. For example, Davis and coworkers[33] have found that for the stearate, oleate, linoleate and linolenate series, the crystal to isotropic liquid transition temperatures are 82,50,42 and 35°C, respectively.

The structures of the liquid crystals formed by cholesteryl esters are not known although it has been shown that some characteristics of the crystal structure are retained[20]. The data for cholesteryl palmitate in Table 2 suggest that formation of the mesophases is accompanied by some trans-gauche isomerisation in the palmitate chain (cf. 34). The packing of the cholesterol residues in cholesterol monohydrate and cholesteryl palmitate crystals is similar (Fig. 4) so that ΔH and ΔS for disordering these regions are likely to be similar in both cases. Thus, to a first approximation, the differences in ΔH and ΔS for melting of cholesterol and cholesteryl palmitate can be attributed to the palmitate chain which is in the β configuration in crystals of the latter compound. This approach suggests that the ΔH and ΔS values associated with disordering the palmitate chain in cholesteryl palmitate are ~ 8 kcal mol^{-1} and 27 cal mol^{-1} deg^{-1}, respectively (this applies to formation of cholesteric and smectic mesophases, or isotropic liquid, because ΔH and ΔS for the transition between these three states are relatively small[31]). The above values are ~ 60% of the equivalent figures for the melting of β crystals of n-hexadecane or palmitic acid suggesting that the palmitate chains in the cholesteric mesophase are more ordered than in isotropic liquid of these simple lipids. Overall, the data in Table 2 suggest that the average fluidity of the palmitate chain in the cholesteric mesophase of cholesteryl palmitate is intermediate to that of the fully hydrated smectic mesophase of dipalmitoyl phosphatidylcholine and the isotropic liquid of palmitic acid. Direct spectral observations of chain motions in cholesterol ester mesophases are required to check this conclusion.

III. INTERACTIONS OF LIPIDS WITH WATER

It is well known that "oil and water don't mix" and the "hydrophobic effect" which give rise to this demixing has been the subject of much study (for reviews, see 35,36). The free energy rise associated with mixing a hydrocarbon chain with water is a consequence of complex changes in water structure which are not well understood. Recently, it has been established[37,38] that the free energy due to hydrophobic interactions for the homologous series of n-alkanes is proportional to the area of apolar surface exposed to the aqueous phase. The hydrophobic interfacial free energy (i.e. the free energy associated with each unit area of hydrocarbon surface in contact with water) is 20-33 cal mol^{-1} $Å^{-2}$. Because of this interfacial free energy, the emulsion formed by dispersing droplets of liquid n-alkane in water is highly unstable and phase separation occurs rapidly. Pure triglyceride and cholesteryl ester oils are similarly immiscible with water because the ester bonds are not sufficiently polar to outweigh the hydrophobic effect.

Increasing the polarity of a lipid causes it to be either water-dispersible of water-soluble. Such molecules are amphi-

philic because they have a polar "head" which mixes with water
and a hydrocarbon chain "tail" which is immiscible with water.
Because of this dual solubility characteristic, amphiphilic
molecules adsorb strongly to interfaces (Fig. 5) (i.e. they are
surface-active). Surfactant molecules can be classified with
respect to their hydrophilic-lipophilic balance on an "H.L.B."
scale. Such values, which increase with the polarity of the
molecule, represent an empirical numerical correlation of the
emulsifying and solubilising properties of different surfactants[39].

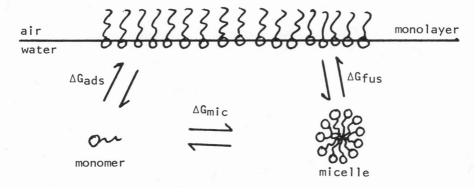

air _____ monolayer
water

ΔG_{ads} ΔG_{fus}

ΔG_{mic}

monomer   micelle

Fig. 5.   Schematic representations of the possible locations of
          hydrophobic lipid molecules when added to water. The
          possible equilibria are shown.

    Fig. 5 summarises the behaviour of lipid molecules which are
sufficiently polar to dissolve in water to give significant concentrations of monomer. Lipids such as soaps have to be heated
above a critical temperature called the Krafft point before they
dissolve. The hydrophobicity of the hydrocarbon chains causes
adsorption at the air-water interface and formation of micelles.
The hydrophobic free energy is proportional to the number of CH_2
groups so the free energies of adsorption (ΔG_{ads}) and micellisation
(ΔG_{mic}) vary linearly with the length of hydrocarbon chain. The
chain-length dependences of these two functions and the free energy
of transfer (ΔG_{trans}) of surfactant from water to n-alkane oil
are given in cal mol^{-1} by the following equations[35].

$$\Delta G_{trans} = k_1 - 850\,n \qquad [2]$$

$$\Delta G_{ads} = k_2 - 625\,n \qquad [3]$$

$$\Delta G_{mic} = k_3 - 700\,n \qquad [4]$$

n is the number of carbon atoms in the hydrocarbon chain and the constants are a function of the polar group. These equations are qualitatively consistent and indicate that the free energy of transfer of a CH_2 group out of contact with water is about $- 700$ cal mol^{-1} in each of the three processes considered; the differences are due to slight variations in the chain interactions when they are in monolayers, micelles and n–alkane oils.

When amphiphilic molecules are dissolved in water they can achieve segregation of their hydrophobic portions from the solvent by self–aggregation to form the micelles depicted in Fig. 5 (ref. 35). Inverted micelles with a polar group core form when the surfactant is dissolved in a non–polar oil. In aqueous systems the formation of micelles is opposed by a repulsive force originating from the polar groups; the repulsion can involve electrostatic and/or hydration effects. These repulsions oppose the attractive force arising from the hydrophobic effect thereby limiting the size to which micelles can grow; for simple amphiphiles containing a single hydrocarbon chain the aggregation number is of the order of 100. As the total concentration of amphiphile is increased, a transition from the predominantly unassociated amphiphile to the micellar state occurs over a narrow critical range of concentration which is called the critical micelle concentration ("cmc"). Raising the concentration of amphiphile above the cmc leads to an increase in the proportion of amphiphile in micelles while the concentration present as monomer remains more or less constant (if the temperature is below the Krafft point, solid surfactant precipitates above the cmc). When the monomer and micelles are in equilibrium, there is rapid exchange of molecules between the two states.

Some effects of variations in the hydrocarbon chains and polar groups are demonstrated by the cmc's listed in Table 3. For an homologous series such as the sodium soaps, log cmc is a linear function of n; the cmc decreases by a factor of ~ 0.3 when n is increased by two[40]. The importance of polar group interactions is demonstrated by the fact that neutral palmitoyl lysophosphatidylcholine forms micelles more easily than the n–hexadecyl sulphate and trimethylammonium molecules which carry a net charge. The much lower cmc for dipalmitoyl phosphatidylcholine as compared to the lyso compound is consistent with the above variation of cmc with n (or a hydrophobic free energy of $- 700$ cal mol^{-1} CH_2 group); this shows that the very low cmc for dipalmitoyl phosphatidylcholine arises from the presence of two chains[41]. In this case, the aggregation number is ~ 3000 and the structures of these micelles

Table 3

The Effect of Variations in Chain Length and Polar Group on the cmc's of Lipids in Water

	Temp $^{\circ}$C	cmc (M)	Ref.
Sodium soap homologues			
decanoate	50	100×10^{-3}	59
dodecanoate	50	24×10^{-3}	59
tetradecanoate	50	7.3×10^{-3}	59
hexadecanoate	50	2.1×10^{-3}	59
n-hexadecyl (C_{16}) surfactants			
trimethylammonium bromide	25	80×10^{-5}	59
sodium 1-sulphate	25	20×10^{-5}	59
lysophosphatidylcholine	25	1×10^{-5}	60
di C_{16} phosphatidylcholine	20	5×10^{-10}	41
cholesterol	25	3×10^{-8}	42

or liposomes have been the subject of much study (see below). Cholesterol, which has a maximum solubility in water of $\sim 5 \times 10^{-6}$M, starts to form rod-like micelles of aggregation number ~ 500 at the exceptionally low cmc of $\sim 3 \times 10^{-8}$M. The free energy of stabilisation of the micelles contains a contribution due to strong intermolecular attractive forces as well as the hydrophobic effect[42].

A simple surfactant such as sodium palmitate needs a free concentration of $\sim 10^{-3}$M to form an adsorbed monolayer (which is soluble), saturate the surface and give the maximum reduction in surface tension; this effect is due to the adsorption of monomer which at the cmc and above is $\sim 10^{-3}$M. Dipalmitoyl phosphatidyl-choline forms an insoluble monolayer which can exist in a variety of phases[3] so, unlike sodium palmitate, the equilibrium between the surface and the substrate is established relatively slowly. This metastability of the monolayer and slow rate of equilibration is presumably associated with the very low concentration of monomer ($\sim 10^{-10}$M) present in this system and low probability of collision with the surface. The fact that at relatively high concentrations of $\sim 1\%$ w/w phosphatidylcholine ($\sim 10^{-2}$M), the surface tension of the air-water interface is reduced significantly suggests that molecules may transfer directly from the micelles (liposomes) to the interface by some fusion process (Fig. 5). This needs further study but transfer of molecules between membranes and liposomes by this process is known to occur.

When water is added to relatively hydrophobic lipids such as phospholipids they swell and disperse in water rather than dissolve as the monomer. Such behaviour is called lyotropic mesomorphism in that the particular phase obtained is function of both temperature and water content. The formation of these structures is a phenomenon which involves both melting and solution[43] and the polarity of the lipid plays a critical role[44]. There is a minimum temperature below which water cannot penetrate into the crystal lattice of polar lipids[45]. At this penetration temperature (T_c, Krafft point), the hydrocarbon chains melt and water diffuses into the polar region of the crystal lattice. Once this occurs the phospholipid crystal can swell in water, and myelin tube formation can proceed spontaneously[45]. The T_c depends upon the nature of the hydrocarbon chains and of the polar region of the molecule, the amount of water present, and on any solutes present in the water. The chain-melting phase transition which occurs in anhydrous lipids also occurs in the hydrated systems so that when the sample is cooled to below T_c, the hydrocarbon chains arrange themselves into an orderly crystalline lattice, but the water is not necessarily expelled from the system. Lyotropic phases containing crystalline paraffin chain regions are called gels (see Fig. 6a); these gels may or may not be metastable. If they are metastable they transform over a period of time to a suspension of microcrystals of the amphiphile in water – the coagel state.

Many phase diagrams, based upon X-ray diffraction experiments carried out with lipid-water samples of different composition as a function of temperature, are now available in the literature (for reviews, see 12,13,17,18,25). Utilising the notation of Luzzati and coworkers[46], the one-dimensional lamellar phases (Fig. 6a,b) are designated L, while the two-dimensional hexagonal structures (Fig. 6c,d) are specified H. In hexagonal I, the cylinders of phospholipid molecules with the hydrocarbon chains to the interior are separated from each other by water, whereas in the inverse hexagonal II phase the cylinders are embedded in a matrix of hydrocarbon chains. Two-dimensional oblique or rectangular lattices are designated P. The homogeneous conformations of the hydrocarbon chains specified by α, β, β′ have been described earlier. A mixture of α and β hydrocarbon chain conformations is designed αβ when the relative content varies with temperature and/or concentration; a phase in which the relative content of the two types of conformation is fixed by the symmetry of the lattice is called γ (ref. 47).

For all the lipids considered in Fig. 7, the bilayer is the predominant structure when more than about 20 wt% water is present. The progressive increase in the lamellar spacings of dimyristoyl and egg phosphatidylcholines with increasing water finally ceases with the formation of two phases when about 45 wt% water is present. The lateral expansion of the bilayer due to the incor-

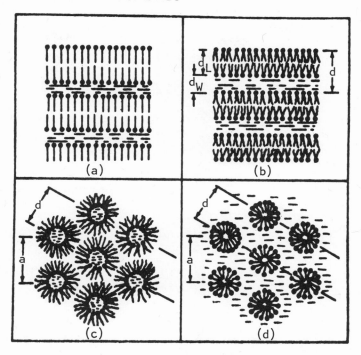

Fig. 6. Schematic representations of lipid—water phases. (a) gel;
         (b) lamellar liquid—crystalline; (c) hexagonal type II;
         (d) hexagonal type I.

poration of water[48] leads to thinning of the bilayer. The phase
behaviour of sphingomyelin—water systems[49,50] parallels that of
phosphatidylcholine—water mixtures; in both cases limited amounts
of water are incorporated into the polar group regions, which
carry no net charge at physiological pH's, of the bimolecular
lamellae. The lamellar phase also predominates in phosphatidylethanolamine—water systems with a hexagonal phase (H II) appearing
only when the temperature is elevated[18]. Unlike phosphatidylcholine, phosphatidylethanolamine is not hygroscopic and does not
swell as strongly when mixed with water; phosphatidylethanolamine
is fully hydrated above T_c when ~ 20 wt% water is added, whereas
excess water only appears with phosphatidylcholine when more than
40 wt% water is present. This increased accommodation of water
can be attributed in part to the greater lateral expansion of
phosphatidylcholine bilayers which arises from differences in the
orientation of the zwitterions in phosphatidylcholine and phosphatidylethanolamine[51]. The net charge on acidic phospholipids leads
to them having dramatically different swelling properties to a

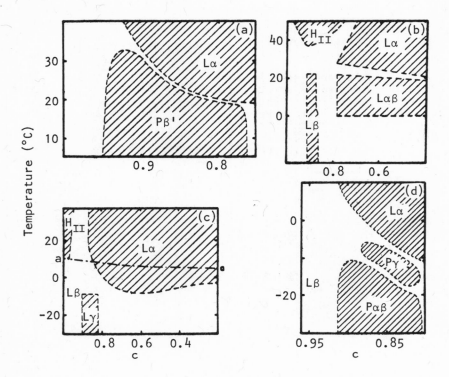

Fig. 7. Portions of the phase diagrams of various lipid-water
systems. The regions where one phase is found pure are
hatched. c is the weight concentration:lipid/(lipid +
water). (a) Dimyristoyl phosphatidylcholine; (b) egg
phosphatidic acid; (c) mitochondrial lipids; (d) egg
phosphatidylcholine (from Ranck et al.[47]).

neutral phospholipid such as phosphatidylcholine. However, as with
phosphatidylcholine, the lamellar phase is predominant; for example,
egg phosphatidic acid forms bilayers exclusively when the water
content is above about 25 wt% (Fig. 7b). Acidic phospholipid
bilayers separate much more than bilayers comprising neutral
phosphatidylcholine molecules. Gulik-Kryziwicki et al.,[52] have
shown that incorporation of as little as about 1% of either cationic or anionic lipid into the lamellar phase in excess water of egg
phosphatidylcholine makes the swelling behaviour comparable to that
of phosphatidylserine. This enhanced swelling arises because of
electrical double layer repulsion between adjacent bilayers[51]. For
this reason, total phospholipid extracts from mammalian cell
membranes swell to incorporate large amounts of water between the
bilayers; for example, beef heart mitochondrial lipids can
incorporate up to 80 wt% water (see Fig. 7c).

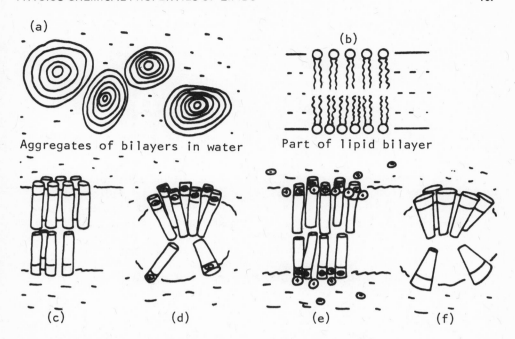

Fig. 8. Effect of shape and charge on aggregation. (a) Phospholipid aggregates present in an aqueous dispersion of phosphatidylcholine. Each concentric line in structure a represents a bimolecular leaflet, as shown in structure b. (c) Bimolecular leaflet formed by cylindrical phospholipid molecules. (d) Structure with increased curvature of the lipid—water interface formed in solution of low ionic strength by phospholipid molecules with net negative charge on the head groups. (e) Curvature of interface of structure d reduced by shielding in the presence of excess ions. (f) Spherical micelle formed by wedge-shaped molecules with or without charge. (From Phillips[3]).

The above brief outline of the phase behaviour of lipid—water systems indicates that the form of the hydrated polymolecular structures depends upon the shape and charge of the lipid molecule. The phosphatidylcholine molecule is roughly cylindrically shaped, allowing easy side-by-side stacking for attainment of the maximum interaction energy. There is no net charge on the polar head group which can give rise to a disrupting, repulsive, electrostatic energy either normal or parallel to the plane of a bilayer and as a result phosphatidylcholines form stable multilamellar aggregates of continuous bimolecular leaflets (Fig. 8a–c). Because they are insufficiently hydrophobic, phosphatidylcholines

with less than 10 carbon atoms in their hydrocarbon chains will
not form stable bimolecular lamellae at ordinary temperatures[3].
If the lipid molecules are ionized, a mutual repulsion between the
polar ends of the molecules is introduced, and there is a tendency
for the bilayer to break up into aggregates of radially oriented
molecules (Fig. 8d). When phosphatidylserine is dispersed in
aqueous solutions of low ionic strength, small bilayer structures
are formed with high curvature of the lipid—water interface,
whereas, in 0.1M sodium chloride solutions, larger multilamellar
aggregates are stable because of shielding of the negative charge
on the phosphatidylserine molecule by counterions (Fig. 8e). This
shielding leads to less curved lipid—water interfaces and less
repulsion between the bilayers. As yet there is no evidence for
spherical micelle formation in acidic diacyl phospholipid—water
systems. When the polar group of the phospholipid has a different
cross section from its apolar hydrocarbon chains (i.e., the
molecule is wedge-shaped) the maximum interaction energy is
obtained with a curved polar group—water interface. When the
polar group cross section is large (e.g. lysolecithin), either
a hexagonal I phase (Fig. 6d) or spherical micelle (Fig. 8f)
is formed depending upon the water concentration. Reduction in the
size of the polar group (e.g., phosphatidylethanolamine) can result
in a hexagonal II phase (Fig. 6c).

IV. MIXING BEHAVIOUR OF LIPOPROTEIN LIPIDS

The mixing properties of the lipids listed in Table 1 are
clearly relevant to lipoprotein structure and we address ourselves to this topic here. Since triglycerides and cholesteryl
esters do not hydrate and are immiscible with water whereas
phospholipids hydrate strongly and swell in water, it is
convenient to consider mixing in terms of these two categories.

Generally, the physical state of the hydrocarbon chains is
crucial in determining miscibility. In anhydrous systems which
are totally liquid, the lipids are completely miscible but cooling
to cause a change of phase of one component can lead to some
immiscibility. For example, cholesteryl linolenate can be incorporated into anhydrous egg phosphatidylcholine mesophases up to
only 25% w/w; the cholesteryl ester molecules are situated in the
hydrocarbon chain regions of the liquid crystal lattice and decrease
the T_c of phosphatidylcholine[53]. Similarly, a feature of the interaction between cholesteryl esters or cholesterol with triglyceride
is that lower melting components depress the melting point of the
high melting ones[54]. Furthermore, triglyceride can remove the
mesophases·in mixtures with cholesterol and cholesteryl linoleate;
e.g., 1–3% triolein removes the cholesteric mesophase and with ﹥50%
no mesophase is observed and crystallisation takes place directly
from the isotropic liquid. The low mutual solid solubility of cholesterol, cholesteryl ester and triglyceride means that crystals

of pure cholesterol separate readily at ordinary temperatures[54].
In order for lipids of a given class to form solid solutions,
their hydrocarbon chains must be of similar length and degree of
unsaturation so that their intermolecular interactions and melting
points are close.

Addition of water can have a dramatic effect on the mixing of
lipids such as cholesteryl esters and phosphatidylcholine. Janiak
et al.,[55] have investigated the egg phosphatidylcholine/cholesteryl
linolenate system and found that addition of water converts phosphatidylcholine into an Lα phase in which cholesteryl linolenate is
soluble to only ~ 2% (i.e. 1 molecule cholesteryl linolenate to 40
phosphatidylcholine molecules). Furthermore, hydration of the
phosphatidylcholine polar groups prevents these molecules entering
the hydrophobic, oily cholesteryl ester phases; phosphatidylcholine
separates as the hydrated Lα phase. Cholesterol can dissolve in
the latter phase at levels up to 33% (i.e. up to equimolar amounts)
and the behaviour of cholesterol/cholesteryl ester/phospholipid
mixtures in excess water is summarised in Fig. 9. The data in
Table 1 indicate that the lipid compositions of serum lipoproteins
fall in Zone III and at equilibrium would form two phases, viz.
phospholipid bilayer containing up to 33% cholesterol and 2%
cholesteryl ester and a liquid or liquid–crystalline cholesteryl
ester phase (liquid triglyceride would dissolve in this phase).
The fact that LDL lipids do not separate in this way in vivo means
that either the existence of LDL particles[34] involves non–equilibrium effects or that the apoproteins perturb the phase behaviour
summarised in Fig. 9.

The solubilisation of the water–immiscible cholesteryl ester/
triglyceride phase into water involves coating the surface of
droplets of this material with hydrated polar lipid, cholesterol
and protein. Thus, the formation of lipoprotein particles involves
preparing a microemulsion of triglyceride/cholesteryl ester in
water. Unless heated[56], such a microemulsion is stable in that
coalescence of the particles and separation of a bulk oil phase
does not occur. Since phospholipids can emulsify triglyceride oils
into water very effectively, the presence of apoprotein as additional
emulsifier seems superfluous, at first glance. However, as shown in
Table 4, the cholesterol acts in an antagonistic fashion to the
phosphatidylcholine and can destabilise triglyceride oil–in–water
emulsions. In order to understand these effects it is necessary
to consider some emulsion science.

The fact that water–dispersible phosphatidylcholine molecules
stabilise oil–in–water (O/W) emulsions whereas the relatively
water–insoluble cholesterol molecules stabilise water–in–oil (W/O)
emulsions is consistent with Bancroft's Rule[57] which states that
the phase in which the emulsifying agent is more soluble will be

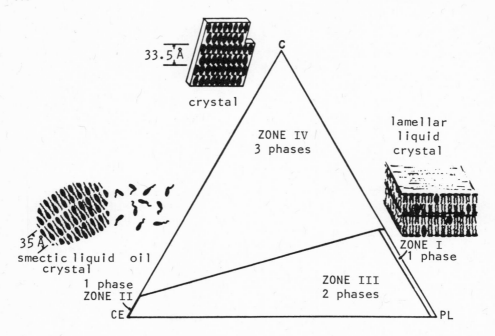

Fig. 9.   Phase diagram of three-component system, cholesterol (C),
          phospholipid (**PL**), cholesteryl ester (**CE**), in excess water
          at 37°C and 1 atmosphere pressure. Zone 1 has a single
          phase of phospholipid lamellar liquid crystal with up to
          33% cholesterol and 2% cholesteryl ester. Zone II also
          has a single phase composed of liquid or liquid crystalline
          cholesteryl ester. Mixtures in zone III have both the
          cholesteryl ester and the phospholipid phase present.
          In zone IV there is a third phase as well, cholesterol
          monohydrate crystals. Schematic molecular representation
          of the phases shown near each apex of the triangle.
          Molecules of cholesteryl ester form isotropic oil
          droplets when random, and birefringent droplets when
          ordered in layers 35 A thick. (From Katz et al.[61])

the continuous one. This result arises because the newly formed
interfaces of the small droplets of the disperse phase produced
during emulsification are not immediately in equilibrium with the
surrounding continuous phase. Due to a relative scarcity of
surfactant in the thin film of continuous liquid between two
juxtaposed droplets, less adsorption occurs in this region and the
interfacial free energies are higher than in other regions of the
droplets which are exposed to the open solution and where diffusion
and adsorption of surfactant can occur freely. The resultant inter-

Table 4

The Effect of Mixtures of Lecithin and Cholesterol on Triglyceride Oil Emulsions[a]

Ratio Lecithin/Cholesterol (w/w)	Emulsion Type
19	O/W
10	O/W
8	W/O/W (duplex)
6	W/O
4	W/O
2	W/O

a. These data for 50% v/v olive oil and water emulsions are taken from Corran[61].

facial tension gradients cause a flow of surfactant and underlying continuous phase into the thin film region so that coalescence is inhibited and the droplets are stabilised. Adsorption from the interior of the dispersed droplet would not result in a surface tension gradient; thus no stabilisation of freshly formed droplets occurs and phase inversion of the emulsion can occur. Such destabilisation of the O/W systems can occur when $\leq 8/1$ w/w lecithin/cholesterol is used as the emulsifier (Table 4). Since the phosphatidylcholine/cholesterol ratios in lipoproteins are less than this value (Table 1), the cholesteryl ester and triglyceride in serum lipoproteins could not be satisfactorily emulsified into water because of the antagonistic effects of phosphatidylcholine and cholesterol. Of course, since the phase volume of fat in serum is only ~ 1% complete inversion is not possible but local destabilisation could occur. The above reasoning therefore suggests that the apoprotein present at the surface of lipoprotein particles is necessary to stabilise the O/W systems.

V. SUMMARY AND CONCLUSIONS

The physical properties of cholesterol, cholesteryl esters, triglycerides, phospholipids and their interactions with water are considered. Hydrocarbon chains adopt the all trans configuration in crystals and lipids can exist in different polymorphic forms because of variations in the angle of chain packing. Trans-gauche isomerisations occur in the chains on heating and a cooperative melting transition takes place at a characteristic temperature; the cooperativity in HDL and LDL particles is low because of their small size. Cholesteryl esters and phospholipids do not only melt to the isotropic liquid but can form liquid crystalline phases in which the molecular order is greater than in the isotropic liquid. Lipids are surface active and associate readily in water because of the hydrophobic effect; the hydrophobic free energy is proportional to the length of the n-alkane chain. The phase behaviour of hydrated lipid systems is a function of the shape and charge of the lipid molecules. While liquid triglycerides and cholesteryl esters mix to form water-immiscible oils, phospholipids tend to form separate structures because the polar groups interact strongly with water. The apoproteins in lipoproteins are probably necessary to stabilise the particles in the aqueous phase because phosphatidylcholine and cholesterol have antagonistic emulsifying actions so that, at the ratios which exist in lipoproteins, they cannot stabilise an oil-in-water emulsion satisfactorily.

REFERENCES

1. R.I. Levy, D.W. Bilheimer and S. Eisenberg, in "Plasma Lipo-proteins", ed. R.M.S. Smellie, Biochem. Soc. Symp. No. 33, Academic Press, London, 1971, pp. 3-17.
2. A.M. Scanu, in "Atherogenesis: Initiating Factors", Elsevier, Amsterdam, 1973, pp. 223-249.
3. M.C. Phillips, Prog. Surface Membrane Sci. 5 (1972) 139-221.
4. A.G. Lee, Prog. Biophys. Mol. Biol. 29 (1975) 3-56.
5. G. Lagaly, Angew. Chem. 15 (1976) 575-586.
6. B.D. Ladbrooke and D. Chapman, Chem. Phys. Lipids 3 (1969) 304-356.
7. J.M. Sturtevant, in "Quantum Statistical Mechanics in the Natural Sciences", ed. B. Kursonoglu, S. Mintz and S. Widmayer, Plenum Press, New York, 1974, pp. 63-84.
8. H.J. Hinz and J.M. Sturtevant, J. Biol. Chem. 247 (1972) 6071-6075.
9. D. Marsh, A. Watts and P.F. Knowles, Biochim. Biophys. Acta 465 (1977) 500-514.
10. D.G. Dervichian, Prog. Biophys. Mol. Biol. 14 (1964) 263-342.
11. D. Chapman, "The Structure of Lipids", Methuen, London, 1965.
12. V. Luzzati, in "Biological Membranes", ed. D. Chapman, Academic Press, New York, 1968, pp. 71-123.
13. R.M. Williams and D. Chapman, Prog. Chem. Fats & Other Lipids 11 (1970) 1-79.
14. A. Tardieu, V. Luzzati and F.C. Reman, J. Mol. Biol. 75 (1973) 711-733.
15. Hitchcock, P.B., Mason, R., Thomas, K.M. and Shipley, G.G. Proc. Nat. Acad. Sci. USA 71 (1974) 3036-3040.
16. Hitchcock, P.B., Mason, R. and Shipley, G.G., J. Mol. Biol. 94 (1975) 297-299.
17. Y.K. Levine, Prog. Biophys. Mol. Biol. 24 (1972) 3-74.
18. G.G. Shipley, in "Biological Membranes Vol. 2", ed. D. Chapman and D.F.H. Wallach, Academic Press, New York, 1973, pp. 1-89.
19. B.M. Craven, Nature 260 (1976) 727-729.
20. B.M. Craven and G.T. DeTitta, J. Chem. Soc. Perkin II (1976) 814-822.

21. K. Van Putte, W. Skoda and M. Petroni, Chem. Phys. Lipids 2 (1968) 361–371.
22. G.W. Gray, "Molecular Structure and the Properties of Liquid Crystals", Academic Press, New York, 1962.
23. D.M. Small, C. Loomis, M. Janiak and G.G. Shipley, in "Liquid Crystals and Ordered Fluids", ed. J.F. Johnson and R.S. Porter, Plenum Press, New York, 1974, Vol. 2, pp. 11–22
24. D. Chapman, Science J. (1965) 32–38.
25. D. Chapman, in "Biological Membranes", ed. D. Chapman, Academic Press, New York, 1968, pp. 125–202.
26. H. Hauser, in "Water. A Comprehensive Treatise", ed. F. Franks, Plenum Press, New York, 1975 Vol. 4, pp. 209–303.
27. M.C. Phillips, R.M. Williams and D. Chapman, Chem. Phys. Lipids 3 (1969) 234–244.
28. J.E. Rothman, J. Theor. Biol. 38 (1973) 1–16.
29. A.G. Lee, N.J.M. Birdsall, Y.K. Levine and J.C. Metcalfe, Biochim. Biophys. Acta 255 (1972) 43–56.
30. A. Bondi, Chem. Rev. 67 (1967) 565–580.
31. E.M. Barrall and J.F. Johnson, in "Liquid Crystals and Plastic Crystals", Vol. 2, ed. G.W. Gray and P.A. Winsor, Horwood, Chichester, 1974, pp. 254–306.
32. D.M. Small, in "Surface Chemistry of Biological Systems", ed. M. Blank, Plenum Press, New York, 1970, pp. 55–83.
33. G.J. Davis, R.S. Porter, J.W. Steiner and D.M. Small, Mol. Cryst. Liquid Cryst. 10 (1970) 331–336.
34. D. Atkinson, R.J. Deckelbaum, D.M. Small, and G.G. Shipley, Proc. Nat. Acad. Sci. USA 74 (1977) 1042–1046.
35. C. Tanford, "The Hydrophobic Effect", Wiley, New York, 1973.
36. F. Franks, in "Water. A Comprehensive Treatise", ed. F. Franks Plenum Press, New York, Vol. 4, 1975, pp. 1–94.
37. R.B. Hermann, J. Phys. Chem. 76 (1972) 2754–2759.
38. J.A. Reynolds, D.B. Gilbert and C. Tanford, Proc. Nat. Acad. Sci. USA 71 (1974) 2925–2927.
39. J.T. Davies and E.K. Rideal, in "Interfacial Phenomena", 2nd ed., Academic Press, London, 1963, p. 371.
40. M.N. Jones, in "Biological Interfaces", Elsevier, Amsterdam, 1975, p. 69.
41. R. Smith and C. Tanford, J. Mol. Biol. 67 (1972) 75–83.
42. M.E. Haberland and J.A. Reynolds, Proc. Nat. Acad. Sci. USA 70 (1973) 2313–2316.
43. D.G. Dervichian, Mol. Cryst. 2 (1966) 55–62.
44. D.M. Small, Fed. Proc. 29 (1970) 1320–1326.
45. A.S.C. Lawrence, Mol. Cryst. Liq. Cryst. 7 (1969) 1–57.
46. V. Luzzati and A. Tardieu, Ann. Rev. Phys. Chem. 25 (1974) 79–94.
47. J.L. Ranck, L. Mateu, D.M. Sadler, A. Tardieu, T. Gulik-Krzywicki and V. Luzzati, J. Mol. Biol. 85 (1974) 249–277.
48. D.M. Small, J. Lipid Res. 8 (1967) 551–557.
49. F. Reiss-Husson, J. Mol. Biol. 25 (1967) 363–381.

50. G.G. Shipley, L.S. Avecilla and D.M. Small, J. Lipid. Res. 15 (1974) 124-131.
51. H. Hauser and M.C. Phillips, Prog. Surface Membrane Sci., in press.
52. T. Gulik-Krzywicki, A. Tardieu and V. Luzzati, Mol. Cryst. Liq. Cryst. 8 (1969) 285-291.
53. C.R. Loomis, M.J. Janiak, D.M. Small and G.G. Shipley, J. Mol. Biol. 86 (1974) 309-324.
54. B. Lundberg, Acta Chem. Scand. B30 (1976) 150-156.
    S. Ekman and B. Lundberg, Acta Chem. Scand. B30 (1976) 825-830.
55. M.J. Janiak, C.R. Loomis, G.G. Shipley and D.M. Small, J. Mol. Biol. 86 (1974) 325-339.
56. R.J. Deckelbaum, G.G. Shipley and D.M. Small, J. Biol. Chem. 252 (1977) 744-754.
57. W.D. Bancroft, J. Phys. Chem. 17 (1913) 501-519.
58. A.E. Bailey, in "Melting and Solidification of Fats", Interscience, New York, 1950, p. 172.
59. P. Mukerjee and K.J. Mysels, "Critical Micelle Concentrations of Aqueous Surfactant Systems", National Bureau of Standards Publications, Washington, 1971.
60. M. Hayashi, M. Okazaki and I. Hara, Proc. 6th Internat. Congress Surface-Activity, Carl Hanser, Munich, Vol. 2, 1973, 361-370.
61. J.W. Corran, in "Emulsion Technology", 2nd ed., Chemical Publishing Co., Brooklyn, 1946, pp. 176-192.
62. S.S. Katz, G.G. Shipley and D.M. Small, J. Clin. Invest. 58 (1976) 200-211.

THE USE OF PARAMAGNETIC PROBES IN THE STRUCTURE DETERMINATION OF HIGH DENSITY SERUM LIPOPROTEINS

H. Hauser

Eidgenössische Technische Hochschule, Zürich

Laboratorium für Biochemie, Universitätstr. 16, CH-8006 Zürich, Switzerland

Introduction

The subject of this paper is the structure of high density lipoproteins (HDL) with particular reference to the nature of the lipid - protein interaction. The question of whether or not there are any general principles underlying the lipid - protein interaction in serum lipoproteins is of prime importance and will be discussed here. The work to be described was carried out with the HDL3 fraction isolated from porcine plasma and characterized by an operationally defined density of 1.12 - 1.21 g/ml. Results obtained with the native system will be compared with those of a model system reconstituted from lecithin and apoproteins isolated from porcine HDL3 lipoprotein. Several papers have discussed the striking compositional and structural similarities of human and porcine lipoproteins (1-4), and it is believed that the principles governing the lipid - protein interaction are generally applicable. To support this some experiments were repeated with human HDL of an operationally defined density d = 1.063 - 1.21 g/ml. The physical techniques used to derive the structural information to be discussed here are mainly NMR and ESR spectroscopy. Because of the small size of the high density lipoprotein particles (see below) the tumbling rate is sufficiently fast as to average out any residual dipolar broadening of NMR lines. Hence good high resolution spectra are obtained from high density lipoproteins.

1. NATIVE HIGH DENSITY LIPOPROTEINS

The structure (morphology) of high-density lipoprotein (HDL_3).
Small-angle X-ray scattering.

The form of the small-angle X-ray scattering curve (Fig. 1, from the work of Atkinson et al. (5)) and the two well-resolved secondary maxima are typical for a highly symmetrical structure and/or a high degree of particle homogeneity. The small-angle region ($h < 0.38$ nm^{-1})

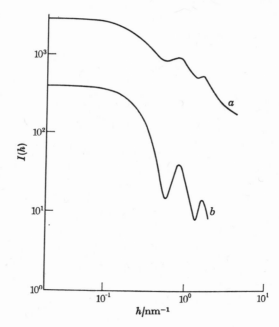

Fig.1. Small-angle X-ray scattering curve from HDL_3 in 0.15 M NaCl, 1 mM EDTA, 0.02% NaN_3 at infinite dilution.
(a) containing the influence of the line collimation system
(b) "desmeared" curve, after correction for these effects. I (h) in arbitrary units (from ref. 5).

was analyzed according to the treatment of Guinier
yielding a radius of gyration R_g = 5.5 nm. The value of
R_g and the position and relative intensities of the sub-
sidiary scattering maxima are not consistent with a
symmetrical particle of uniform electron density. From
a comparison of the experimental scattering curve with
theoretical curves computed for models consisting of a
lipid core of low electron density and an electron dense
outer shell the following conclusions were derived (5):
HDL_3 is a spherical particle with a radius of gyration
of 5.5 nm consisting of a central, low electron density
(lipid) core of radius 4.2 nm and an outer electron
dense, polar layer of a thickness of about 1.2 nm. The
thickness of the electron dense outer shell of about
1 nm is consistent with the length of the polar group
of phosphatidylcholine or a typical diameter of a protein α-helix suggesting that the outer layer may consist of phospholipid polar groups and protein α-helices.
These features are summarized in Fig. 2. The radius of
gyration of 5.5 nm is in good agreement with the Stokes
radius of 5.6 nm determined by gel filtration on Sepharose 4B and with the apparent radius of 5.3 nm determined from electron micrographs of negatively stained
HDL_3 samples.

The X-ray scattering at wider angles showed a single
broad peak at an equivalent Bragg spacing of 0.45 nm
characteristic of highly mobile, liquid crystalline
hydrocarbon chains (5).

NMR results (6,11)

Fig. 3a and b are 270 MHz ^{1}H-NMR spectra of a sonicated egg lecithin dispersion and of HDL_3, respectively
(6). From a comparison of the two spectra it is clear
that (1) the lipoprotein spectrum is essentially a lecithin spectrum and (2) that, apart from a few resonances
at 7.2 ppm (from 3-(trimethylsilyl)propane sulphonate)
arising from aromatic protons and probably at 2.2 ppm
no protein signals are visible. 360 MHz ^{1}H-NMR spectra
of human HDL (d = 1.063 - 1.21 g/ml) closely resemble
the spectrum shown in Fig. 3b (Hauser, H. and Kostner,
G., to be published). Similar to the ^{1}H-NMR spectra
most of the prominent resonances of the ^{13}C-spectra of
both porcine HDL_3 and human HDL can be assigned to the

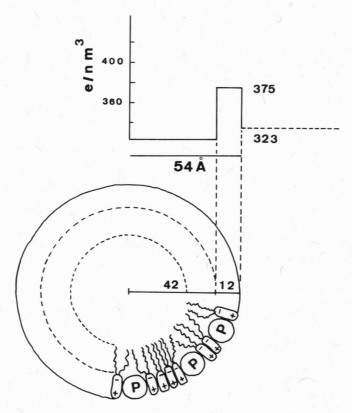

Fig. 2. Summary of the small-angle X-ray scattering data
of HDL₃. The spherical particle of a radius of 5.4 nm
is surrounded by an electron dense layer (375 e/nm³) of
1.2 nm consisting of the phospholipid polar groups and
the protein (P) a large proportion of which is present
as amphipathic α-helical segments (shown in cross section (P)). The axes of the α-helical segments intercalated between the lipid polar groups run approx. perpendicular to the phospholipid hydrocarbon chains. The
core of the particle (4.2 nm) consists of lipids, mainly
cholesteryl esters, having a low electron density of
323 e/nm³.

Any paramagnetic ion has a local magnetic field

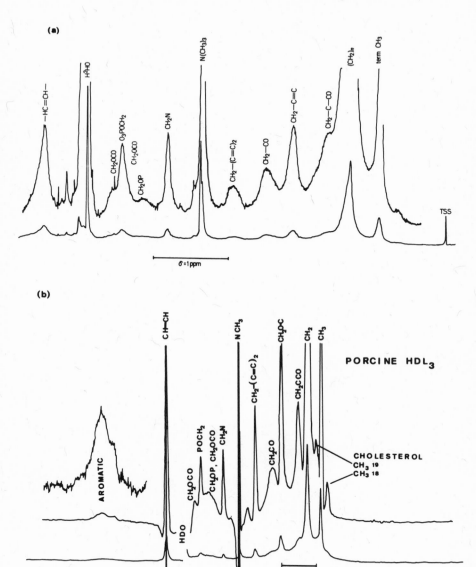

Fig. 3. (a) 270 MHz proton n.m.r. spectrum of a sonicated egg lecithin dispersion (20 mg/ml ≃ 0.027 M) in ^2H_2O (nominal pH 5.5). (b) 270 MHz proton n.m.r. spectrum of native porcine high density lipoprotein HDL_3 (d = 1.12-1.21 g/cm^3) in ^2H_2O (concentration 100 mg/ml) containing 0.15 M NaCl, 1 mM EDTA and 0.02% NaN_3 adjusted to a nominal pH 8.6.

lipid moiety (cf. 14).

The lack of protein resonances indicates that the protein is rather rigid having a high degree of secondary and/or tertiary structure. The NMR results are consistent with CD data showing that ~70% of the protein in HDL_3 is α-helical (9).

Chemical shift measurements (relative to trimethyl-silyl-propane sulphonate (TSS) as an internal standard) on both porcine HDL_3 and human HDL showed that the resonances from the glycerophosphorylcholine group and the first few CH_2 groups of the hydrocarbon chains are shifted relative to the chemical shifts of these resonances in pure egg lecithin. Intensity measurements show that all the $N(CH_3)_3$ groups contribute to the ^{1}H spectrum.

Paramagnetic ions as structural probes

Any paramagnetic ion has a local magnetic field around its centre which can induce considerable perturbation in both chemical shifts and relaxation times of nearby nuclei. The paramagnetic ions used in structural (conformational) studies may often be classified as "shift probe" or "relaxation probe" depending on the type of perturbation induced. The former type of probe induces mainly changes in chemical shifts although the shifted resonances usually show some line broadening, the latter affects primarily the relaxation behaviour and causes line broadening. Here we use the paramagnetic anion $Fe(CN)_6^{3-}$ as a shift probe and the paramagnetic lanthanide Gd^{3+} as a relaxation probe to derive structural information on both native high density lipoproteins and on a reconstituted model system. Gd^{3+} is an effective relaxation probe because the cation has a symmetric electron distribution (f^7) which ensures that effects on chemical shift are negligible.

The addition of paramagnetic $K_3Fe(CN)_6$ to porcine HDL_3 and human HDL produced changes $\Delta\delta$ in the [1]H-chemical shifts of the polar group resonances. 100% of the $N(CH_3)_3$ signal was shifted upfield indicating that all the $N(CH_3)_3$ groups (including those of sphingomyelin) are present at the surface of the HDL-particle readily accessible to the $Fe(CN)_6^{3-}$ anion. Fig. 4 shows the shift changes $\Delta\delta$ of the $N(CH_3)_3$ signal of human HDL as a function of the molar ratio $K_3Fe(CN)_6$ / phospholipid (Hauser, H. and Kostner, G., to be published). The shift

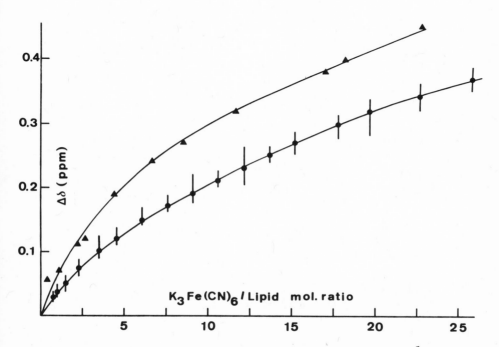

Fig. 4. Changes in chemical shift ($\Delta\delta$) of the ^1H $N(CH_3)_3$ signal as a function of the molar ratio $K_3Fe(CN)_6$ / lipid.
(▲—▲) human HDL (d= 1.063-1.21 g/ml)
(●—●) sonicated egg lecithin dispersion or sonicated phospholipid dispersion (composition: 70% egg lecithin, 20% sphingomyelin and 10% egg lysolecithin). Data were obtained at 360 MHz. The effective lipid concentration, i.e. the concentration of lipid effectively in contact with the $Fe(CN)_6^{-3}$ was 0.0176 M. In case of HDL the effective lipid concentration is identical to the total lipid concentration, in the case of the lipid dispersion the total concentration is 0.0267 M.

changes of the $N(CH_3)_3$ signal of sonicated lecithin and
mixed phospholipid dispersions are shown for comparison.
The changes in chemical shift $\Delta\delta$ of the other polar
group signals were in the order $N(CH_3) > CH_2N > POCH_2 >$
(choline) $> CH_2OP$ (glycerol). No changes in chemical
shift of the [1]H-signals from the lipid hydrocarbon
chains and of the two [1]H-signals from cholesteryl ester
and/or free cholesterol were observed up to $K_3Fe(CN)_6$
concentrations of 0.7M. These results are consistent
with the schematic drawing in Fig. 2 showing that all
phospholipids are arranged as a monomolecular surface
layer surrounding the hydrophobic core. The location
of the cholesteryl esters and free cholesterol is not
conclusive. To solve this problem we intend to use the
spectral perturbation in the ^{13}C-NMR spectrum induced
by paramagnetic lanthanides.

The difference in the shift changes $\Delta\delta$ between lipoprotein and lipid dispersions (Fig. 4) is attributed to
the presence of protein in the surface of the lipoprotein particle affecting the binding and/or accessibility
of the $Fe(CN)_6^{3-}$ anion (diameter approx. 0.9 nm). The
surface location of the protein is also apparent from
the line width of the shifted $N(CH_3)_3$ resonance. In the
presence of $K_3Fe(CN)_6$ the $N(CH_3)_3$ signal of HDL was
broadened by a factor of 3-4 (at 360 MHz) compared to
the $N(CH_3)_3$ signal in lecithin. The field dependence
of the width (cf. 11) indicates that the broadening is
due to a spread in chemical shifts, i.e. the $N(CH_3)_3$
signals are not equally shifted by $Fe(CN)_6^{3-}$. This reflects differences in the chemical environment and / or
accessibility to $Fe(CN)_6^{3-}$ of the $N(CH_3)_3$ groups arising from the orientation of the protein in the surface
of the lipoprotein particle. Exchange of lipid molecules
between these different environments is slow on the
NMR time scale (less than 10/s). Such slow exchange is
consistent with the protein being intercalated between
patches of lipid monolayer (Fig. 2) giving rise to a
mosaic surface structure of the lipoprotein particle
as discussed before (6). That the accessibility of the
phospholipid polar group extends to cations is evident
from ^{31}P-NMR experiments. The ^{31}P-NMR spectra of human
HDL consist of three chemically shifted signals which,
on the basis of chemical shift and intensity measurements can be assigned to lecithin, lysolecithin and
sphingomyelin, respectively (cf. 15). Within the error
of the measurement the spectral intensities were in

agreement with the phospholipid composition. Adding increasing quantities of $Gd(NO_3)_3$ caused broadening of all three resonances whereby the total signal intensities were affected. This is consistent with the $Fe(CN)_6^{3-}$ experiments indicating that all the phospholipids are arranged in a monolayer on the outer surface of the lipoprotein particle (Fig. 2).

Comparing line width and T_1 measurements of porcine HDL_3 (11) with corresponding measurements on lecithin bilayers it can be concluded that the presence of the protein does not change significantly the molecular motion of the lipid except that the anisotropy of motion is increased (Hauser, H. and Kostner, G., to be published).

2. MODEL SYSTEM

The results obtained with the native HDL_3 are compared with a model system consisting of dimyristoyl-L-3-lecithin and apoprotein extracted from porcine HDL_3. The lipoprotein (lipid/protein molar (weight) ratio = 100 (2.5)) was prepared by mixing the unsonicated lipid dispersion with the appropriate amount of protein and subjecting the mixture to short bursts of ultrasonication. The above molar ratio represents the minimum amount of lipid required to completely complex one apoprotein molecule as determined by density gradient centrifugation (10).

Small-angle X-ray scattering (7) shows that the reconstituted system differs from the spherical geometry of the native HDL_3 system. The complex is an oblate ellipsoid with 11.0 nm as the major axis and 5.5 nm as the minor axis. The oblate shape is dictated by the fundamental bilayer organisation of the lecithin being retained (see discussion below). Consistent with that is the observation of a sharp diffraction at an equivalent Bragg spacing of 0.42 nm at temperatures below 30ºC and the transition of this diffraction to a broad scattering peak at an equivalent Bragg spacing of 0.45 nm at temperatures > 30ºC. Such a transition is characteristic of the crystal-to-liquid crystal transition in phospholipid bilayers. Using synthetic lecithins differing in hydrocarbon chain length it can be shown that the dimension of the minor axis correlates with the length of the hy-

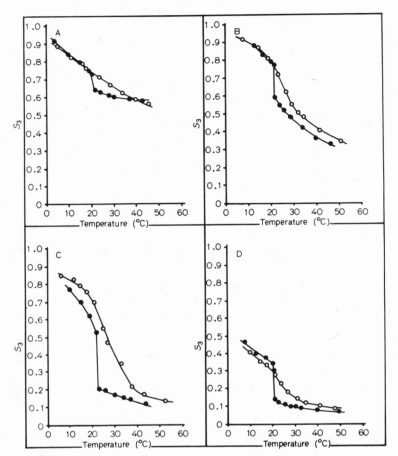

Fig. 5. Order parameters (S_3) as a function of temperature. Dimyristoyl lecithin dispersions (–●–) and the
model lipoprotein (–o–) reconstituted from dimyristoyl
lecithin and apoprotein extracted from porcine HDL_3
were spin-labelled with the following fatty acid spin
labels I (m,n);

$$CH_3 (CH_2)_m - C - (CH_2)_n COOH$$

(a) I (12,3) pH 7.5 (5 mM Tris, 1 mM EDTA, 50 mM NaCl,
    0.02% NaN_3)
(b) as (a) but pH 3.5 adjusted by the addition of HCl.
(c) I (5.10) pH 7.5
(d) I (1,14) pH 7.5
From ref. 8.

drocarbon chains which are oriented perpendicular to the
interface. The difference in the morphology between native and reconstituted system is probably due to the absence of cholesteryl esters in the model. Further evidence for the phospholipid being present as a bilayer is
furnished independently by ESR spin label experiments
and differential scanning calorimetry. Fig. 5 shows that
all spin probes incorporated in dimyristoyl lecithin bilayers report sharp chain-melting transitions between
20 and 23°C. With pure lecithin bilayers at temperatures
above the chain-melting transition the order parameters
as defined in ref. (12) showed a steady decrease with
increasing distance of the nitroxide radical from the
lipid - water interface. This so-called flexibility gradient typical for phospholipid bilayers is shown in
Fig. 6. In the reconstituted system the labels with the
nitroxide group close to the centre of the bilayer showed
a broadened phase transition while with the fatty acid
having the nitroxide in position 5 no chain-melting
transition was observed indicating that the prime site
of interaction of the apoprotein is the glycerylphosphorylcholine group of lecithin. Furthermore, all three
fatty acid spin labels used showed higher order parameters in the complex than in the pure bilayer although
the shape of the flexibility gradient in the reconstituted lipoprotein system was identical to that measured
with pure lecithin bilayers (Fig. 6). This indicates
that the lipid is not immobilized as a result of the interaction with the apoprotein and therefore strong interactions between the lipid and the protein can be ruled out. In the complex the lipid motion is more aniso-
tropic as indicated by the higher order parameter. Consistent with the ESR spin label results are differential
scanning calorimetry data showing a main endothermic
peak at 23°C typical for dimyristoyl lecithin bilayers.

The NMR results obtained with the reconstituted system (cf. 8) are consistent with the data reported for
the native lipoproteins indicating that the mode of the
lipid-protein interaction is similar in the two systems.
This is remarkable considering the difference in the
morphology of the two systems.

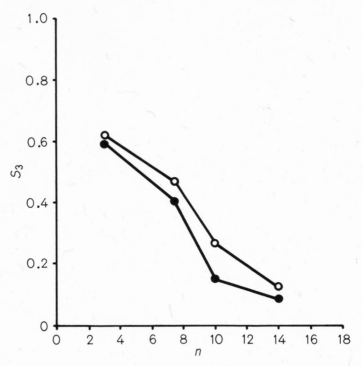

Fig. 6. Order parameter S_3 at $35^{\circ}C$ as a function of the position of the nitroxide group along the hydrocarbon chain; dimyristoyl lecithin bilayers (—●—); lipoprotein complex reconstituted from dimyristoyl lecithin and porcine HDL_3-apoprotein (—o—). From ref. 8.

3. SUMMARY

Comparing the native HDL with the reconstituted system it is concluded that despite morphological differences the principles governing the lipid-apoprotein interaction are similar in both systems. The lipid-protein interaction is characterized as follows:

(1) The prime site of the relatively weak lipid-protein interaction is in the polar region of the phospholipid, i.e. the glycerylphosphorylcholine group.

(2) The surface of the HDL-particle has a mosaic structure consisting of patches of phospholipid monolayer separated by protein. All phospholipids are oriented at the external surface as shown in Fig. 2 with their polar

groups readily accessible to ions in the aqueous phase.

(3) The lipid molecules are not immobilized by the interaction with the protein and therefore strong binding does not occur. The molecular motion of the lipid is similar to that observed in bilayers except that the interaction with the protein increases the anisotropy of motion.

(4) The NMR data are consistent with the protein being fairly rigid and most of the peptide segments being oriented at the lipid-water interface. Since the α-helix content is known to be high the axes of the α-helices must be running along the interface(perpendicular to the hydrocarbon chain axis).The high affinity of the protein for the lipid-water interface discussed before (13) and the specific interfacial arrangement of the peptide segments are probably responsible for the emulsifying properties of apoproteins.

References

(1) Davis, M.A.F., Henry, R. and Leslie, R.B. (1974) Comp. Biochem. Physiol., Vol. 47 B, pp 831-849.

(2) Jackson, R.L., Baker, H.N., Taunton, O.D., Smith, L.C., Garner, C.W. and Gotto, A.M. (1973) J. Biol. Chem. 248, 2639-2644.

(3) Mills, G.L. and Taylaur, C.E. (1971)   Comp. Biochem. Physiol. 40 B, 289-301.

(4) Fidge, N. (1973) Biochim. Biophys. Acta 295, 258-273.

(5) Atkinson, D., Davis, M.A.F. and Leslie, R.B. (1974) Proc. R. Soc. Lond. B 186, 165-180.

(6) Hauser, H. (1975) Febs Letters 60, 71-75.

(7) Atkinson, D., Smith, H.M., Dickson, J. and Austin, J.P. (1976) Eur. J. Biochem. 64, 541-547.

(8) Andrews, A.L., Atkinson, D., Barratt, M.D., Finer, E.G., Hauser, H., Henry, R., Leslie, R.B., Owens, N.L., Phillips, M.C. and Robertson, R.N. (1976) Eur. J. Biochem. 64, 549-563.

(9)   Morrisett, J.D., Jackson, R.L. and Gotto, A.M., Jr.
      (1975) Ann. Rev. Biochem. 44, 183-207.

(10) Hauser, H.,Henry, R., Leslie, R.B. and Stubbs, J.M.
      (1974) Eur. J. Biochem. 48, 583-594.

(11) Finer, E.G., Henry, R., Leslie, R.B. and Robertson,
      R.N. (1975) Biochim. Biophys. Acta 380, 320-337.

(12) Hubbel, W.L. and McConnell, H.M. (1971) J. Amer.
      Chem. Soc. 93, 314-326.

(13) Davis, M.A.F., Hauser, H., Leslie, R.B. and Phillips,
      M.C. (1973) Biochim. Biophys. Acta 317, 214-218.

(14) Hamilton, J.A., Talkowski, C., Childers, R.F., Wil-
      liams, E., Allerhand, A. and Cordes, E.H. (1974)
      J. Biol. Chem. 249, 4872-4878.

(15) Assmann, G., Sokoloski, E.A. and Brewer, Jr., H.B.
      (1974) Proc.Nat.Acad.Sci. USA 71, 549-553.

REASSEMBLY OF APOPROTEINS AND LIPIDS

Maryvonne ROSSENEU

Simon Stevin Instituut voor Wetenschappelijk Onderzoek

Ziekenhuis Sint Jan, Ruddershove, B-8000 Brugge, Belgium

INTRODUCTION

An approach to the problem of lipid-protein interactions in lipoproteins, is to study the behaviour of "in vitro" reassembled apoprotein-lipid mixtures. Apoproteins obtained by delipidation of lipoproteins can be recombined with lipids either synthetic or isolated from the corresponding lipoprotein fraction. Analysis of the kinetics and thermodynamic features of the association process, as well as of the composition, properties and structure of the complex, should provide information about the nature and extent of the forces responsible for the formation and stabilization of the apoprotein-lipid bond within native lipoproteins. Various technical problems are associated with these experiments such as the preparation and characterization of lipid dispersions and of apoprotein solutions, the selection of an optimal reassembly procedure given the lipid and protein composition of the initial mixture. Discussion and evaluation of the various reassembly techniques will be the first part of this paper. The physical techniques applied to the monitoring of the extent of association and to the isolation and characterization of the lipid-apoprotein complexes will be subsequently described. Finally, some of the results obtained with isolated apoA-I, apoA-II, apoC-I, apoC-III proteins will be summarized and analysed.

I. PREPARATION OF THE LIPIDS FOR REASSEMBLY EXPERIMENTS

A. Dispersion of phospholipids

It seems now well established by many "in vitro" recombination studies, specially by LUX et al. (1), SCANU et al. (2,3), that binding of phospholipids is a prerequisite to the binding of neutral lipids to apoproteins. According to HAMILTON (4) nascent HDL appear as an apoA-II-phospholipid complex. This stresses and justifies the importance of the efforts to isolate and characterise phospholipid-apoprotein complexes.

The procedures to recombine apoproteins with phospholipids are dependent on the nature of the latter components. Among the phospholipids a first distinction should therefore be made between micellar phospholipids consisting of lysolecithins and short-chain lecithins, and apolar phospholipids consisting of long-chain lecithins and sphingomyelin (Fig. 1.).

1. Micellar phospholipids. Due to their amphipathic nature phospholipids have a tendency to self-associate into condensed molecular aggregates in aqueous media; in the case of lysolecithins and short-chain lecithins these aggregates are micelles, characterized by their aggregation number, critical micellar concentration, size and shape (5). These properties are determined by the length and degree of unsaturation of the acyl chains, ionic strength, etc. Lysolecithin and short-chain lecithins such as dihexanoyl, dioctanoyl, -L-alpha lecithins are water soluble and form optically clear solutions in water. They can also by used in the monomeric state since their critical micellar concentration is low (6,7). However, as short-chain lecithins are not present in lipoproteins and lysolecithin represents only a minor phospholipid constituent, these binding studies might not be significantly related to the biological conditions.

2. Lecithin and Sphingomyelin. Long-chain lecithins and sphingomyelin require special procedures for their preparation and isolation since they can exist in many physical forms (8,9). Anhydrous preparations of lecithins exhibit various physical states of aggregation as water is added. Such behaviour is called: lyotropic mesomorphism (10,11). When the content of water exceeds about 50% a smectic phase with a liquid-like hydrocarbon region appears. In the presence of an excess of water which is the case in all reassembly experiments carried out so far, lecithin and sphingomyelin associate to form molecular structures or liposomes, whose lamellar configuration is referred to as a "smectic mesophase". This structure can be visualized as stacked lamellae of phospholipid molecules in a bilayer configuration with the water taking up the space between the lamellae and interacting with the polar groups of phospholipids (12, 13). The tendency of phospholipids to assume this hydrated lamellar configuration is related to the temperature for the melting of hydrocarbon chains. Thus, natural unsaturated phospholipids, above this transition at ambient temperature, will form these structures spontaneously. Synthetic phospholipids with saturated acyl chains will also form similar structures if the temperature is raised close to the transition point of the dry compounds (14).

The original preparation of these multilamellar vesicles (15) was made by simple mechanical shaking of a dry phospholipid sample with a salt solution in a round-bottom flask. This method yields heterogeneous mixture of particles of varying size.

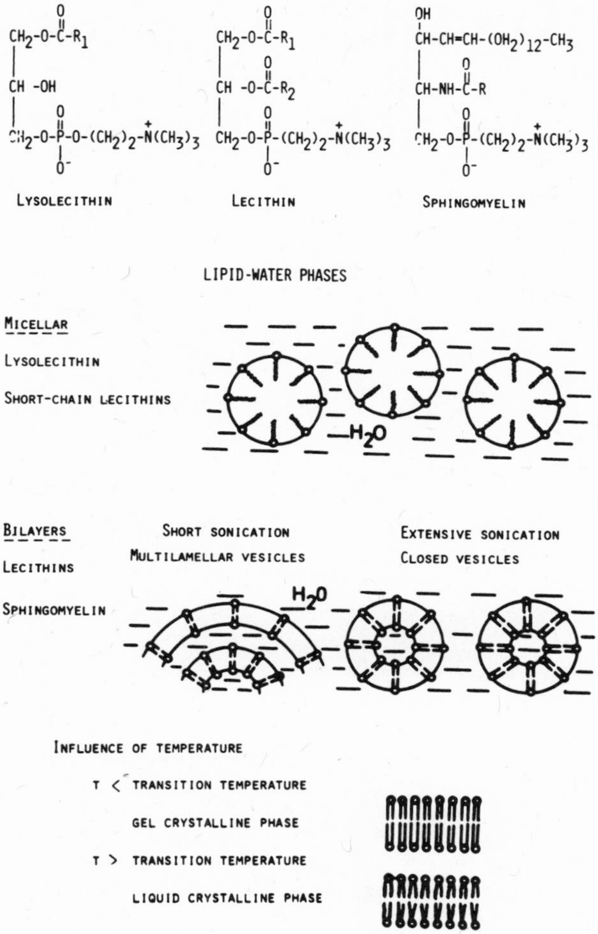

Fig. 1. Physical state of phospholipid dispersions

Exposure of the coarse phospholipid suspensions to ultrasonic radiation tends to break the large multilamellar vesicles into smaller ones and eventually produces a homogeneous population of unilamellar vesicles of 200-500Å diameter (16). Decrease of the size of the vesicles is accompanied by a progressive decrease of the turbidity of the phospholipid dispersion. This can be monitored by measuring the optical density at 450nm (Fig. 2.). After sonication at 37° and 50W of a dimyristoyl lecithin suspension for 15' the optical density remains constant at 0.1. The length of the sonication period depends on the sonication temperature and power. At this stage the suspension consists of more than 80% unilamellar lecithin vesicles.

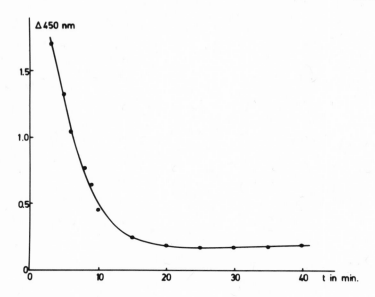

Fig. 2.   Monitoring of the optical density decrease
          during sonication of dimyristoyl lecithin (37°, 50W)

     Light scattering and sedimentation velocity studies of aqueous
dispersions of egg lecithin vesicles indicate an aggregate weight of
1.5 to $2.1.10^6$ daltons, corresponding to approximately 1.9 to $2.7.10^3$
molecules per particle (16). Dimyristoyl lecithin and sphingomyelin
vesicles can be fractionated by gel chromatography on a Sepharose
6B column into large vesicles eluting in the void volume and into
closed single-shelled liposomes (Fig.3). According to the calibra-
tion of this column the molecular weight of the lecithin liposomes
is $1.6.10^6$ and that of sphingomyelin is $3.0.10^6$ daltons.
     The stability of these lipid aggregates is dependent on the
method of preparation, time, temperature and chemical composition.
According to SUURSKUSK et al. the transition temperature of small
vesicles is lower than that of larger vesicles (17). As a function
of time these vesicles tend to coalesce to form larger aggregates.
To prevent this phenomenon they should preferably be kept at temperatures higher than their transition temperature.
     We compared the enthalpy liberated on incubating small and
large dimyristoyl lecithin vesicles with apoHDL at 25°, i.e. above
the phospholipid transition temperature (t = 23°). The complex
formation is more highly exothermal with small vesicles (7). The
enthalpy is a function of both the temperature at which vesicles
sonication occurs and apoprotein-vesicles reassembly takes place.
At t > transition temperature, the enthalpy is not sensitive to the
sonication temperature as phospholipids are in a fluid state. At
a reassembly temperature < transition temperature, the energy liberated during the complex formation increases with the sonication
temperature i.e. with the energetical state of the phospholipid
dispersion (Fig. 4).

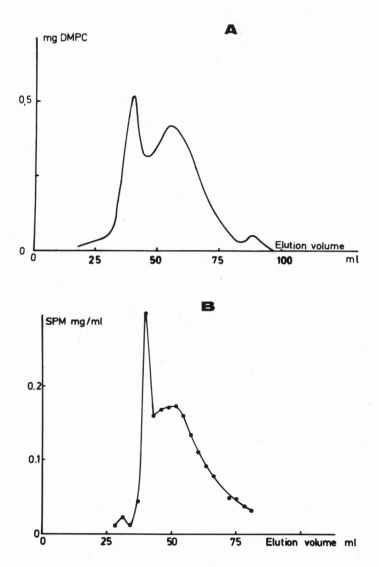

Fig. 3. Fractionation of lecithin (3A) and sphingomyelin
(3B) liposomes on a Sepharose 6B column.

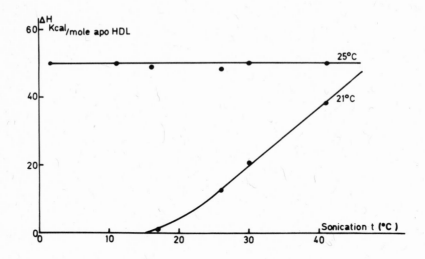

Fig. 4.   Enthalpy change on binding dimyristoyl lecithin
          vesicles to human apoHDL. Reassembly temperatures
          were 25° and 21°, and the lecithin was sonicated
          at temperatures varying between 4 and 40°.

          PHILLIPS et al. (18) have observed that the kinetics of complex
formation between porcine apoHDL and multilamellar lecithin vesicles
are slower than with single-shelled vesicles, as the protein interacts only with the outer bilayer. Eventually the same type of complex was obtained. Therefore closed unilamellar vesicles seem superior to the multilamellar ones for reassembly experiments with maximum yield and reproducibility of complex formation.
          Another procedure for the preparation of unilamellar vesicles
was proposed by BAZRI and KORN (19). It consists in flushing an
ethanolic solution of the phospholipid into the aqueous phase by
means of a microsyringe. The ethanol should be dialysed out of the
vesicle dispersion afterwards and the vesicles concentrated on an
Amicon X-100 membrane. Though this procedure yields small vesicles
it is less suitable for calorimetric experiments as complete dialysis of the ethanol proved difficult and hard to reproduce. As a
consequence the dilution heat of the vesicles was large and the presence of the organic solvent might interfere with protein binding.
BRUNNER et al. (20) have recently proposed a method to prepare single-shelled lecithin-vesicles of uniform size by solubilizing unsonicated lecithin dispersions with sodium cholate and removing the
detergent from the mixed lecithin-cholate micelles by gel filtration
on Sephadex G-50. A homogeneous population of single-bilayer vesicles free of multilamellar structure is obtained by this method.

B. Dispersion of neutral lipids

At the present, there are no methods for preparing pure, homogeneous, triglyceride, cholesterol or cholesteryl ester particles for studying lipid-protein association. However when mixed with lecithin these neutral lipids can exist in well defined uniform structures. Cholesterol-lecithin form stable complexes up to a ratio of 2:1 (21). According to LUNDBERT and SAARINEN (22), stable lecithin-cholesterylester mixtures are obtained by cosonication at a minimal ratio of 0.5/1 (W:W). Fig. 5. shows that, after 30' cosonication at 40°, dimyristoyl lecithin and cholesterol oleate, above a 0.5/1 ratio exist as a stable dispersion with an optical density of 0.1.

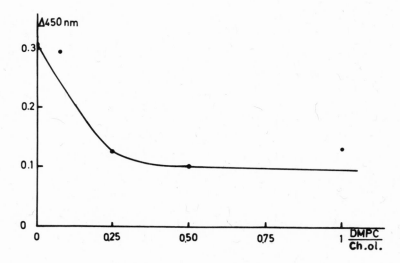

Fig. 5.   Optical density of a dimyristoyl lecithin-cholesterol oleate dispersion after cosonication at 40°, 50W for 30'.

II. APOPROTEIN ISOLATION AND SELF-ASSOCIATION

Procedures for the isolation and purification of apolipoproteins have been extensively described (23, 24, 25). As efficient isolation procedures imply the use of either detergents, urea or guadidium hydrochloride, this suggests that the polypeptides easily self-associate in aqueous buffer solutions. These processes are now well documented for the apoA-I and apoA-II proteins of human and non-human primates (26, 27, 28, 29, 30), suggesting that the lower polymers have a higher affinity for phospholipids (31).

Several agents are able to counteract the self-association process of the apoA-I protein, i.e. detergents such as sodium decyl- or dodecylsulfate (32), lysolecithin (31) or the presence of another apoprotein (33).

We observed that, after preincubation with lysolecithin, the enthalpy change evolved on binding the human apoA-I protein to di-myristoyl lecithin liposomes increases by -25Kcal/mole apoprotein. This difference corresponds to the enthalpy for dissociation and conformational changes of the apoA-I protein (31).

As a consequence, lipid-apoprotein reassembly studies which involve the presence of both lipid and protein aggregate should be performed under carefully controled conditions so as to yield re-producible results, which can be compared to those from other stu-dies.

III. PROCEDURES FOR REASSEMBLING LIPIDS AND APOPROTEINS

Lipid-apoprotein complexes "in vitro" in bulk aqueous solutions have been obtained through two main types of procedure, the one re-quiring ultrasonic irradiation of the phospholipids only, while the other required an extra co-sonication step of the mixture. Both procedures will be described briefly.

A. Procedures with cosonication

As proposed by SCANU (34) a phospholipid dispersion in an aqueous buffer, is sonically irradiated in the presence of nitrogen for a fixed period of time, at a temperature above the melting point of the phospholipid acyl chains, in a container of fixed geometry. After addition of the purified protein, the mixture is further so-nicated. Several authors followed this experimental procedure in earlier reassembly work (35).

We have stepwise reassembled the lipid and protein moiety of HDL, LDL and VLDL by cosonication in the following sequence : phospholipids, plus cholesterol, plus cholesterol esters, plus tri-glycerides, plus apoproteins. Each addition involved a 15' sonica-tion step at 40°.

A comparison of the microviscosity of the reassembled mixture with that of native VLDL and HDL (Fig. 6A and B) shows that the values are very close, suggesting that the reassembly process yields an homogeneous mixture (36). For LDL the microviscosity of the com-plex remains lower, probably due to an incomplete solubilization of apoB (Fig. 6C).

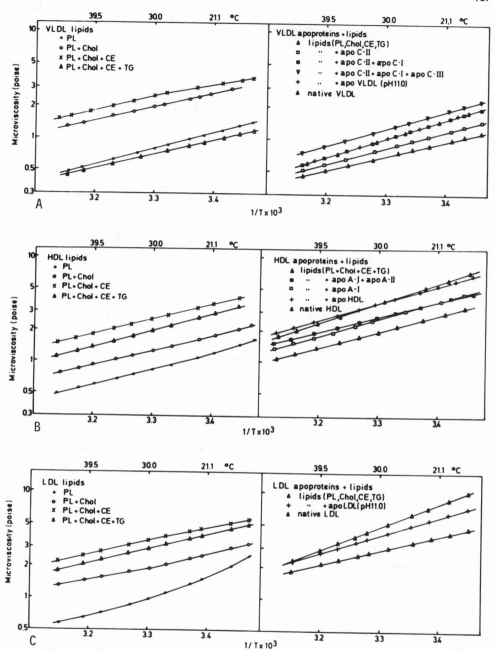

Fig. 6. Microviscosity of stepwise reassembled lipid
and lipid plus apoprotein dispersions.
A : VLDL; B : HDL; C : LDL.

B. Procedures with incubation

    After sonication of the aqueous phospholipid dispersion, the
protein can be subsequently incubated with the lipids without sonication at a temperature above or around the transition point. This
procedure is specially useful for monitoring either the kinetics of
the reaction by spectrophotometry or fluorescence (37), or the heat
effect by mixing calorimetry (38). It also avoids any possible denaturation or conformational changes of the protein due to sonic
irradiation. Moreover as phospholipid-apoprotein binding are rapid,
exothermal processes, the initiation of the reaction does not require any additional energy. This procedure might not be applicable
to lipid mixtures containing cholesteryl esters or triglycerides.

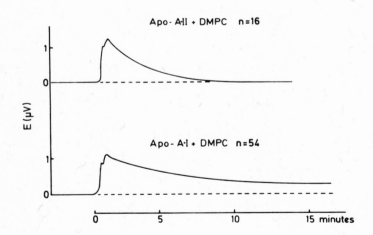

    Fig. 7A.   Thermograms for the binding of dimyristoyl
               lecithin liposomes to human apoA-I and
               apoA-II proteins
               n = mole lecithin/mole apoprotein

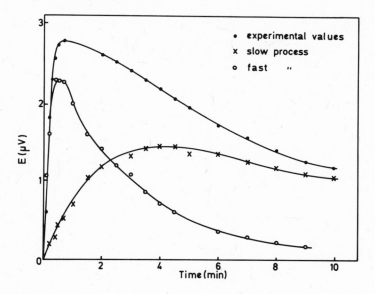

Fig. 7B.   Analysis of the above thermogram into
           two components.

APO A-I - DMPC ASSOCIATION

NATIVE HUMAN APOA-I

N < 100   APoA-I$_P$ + DMPC $\xrightarrow{\text{FAST}}$ APoA-I$_D$ $\xrightarrow{\text{FAST}}$ APoA-I$_D$-DMPC $\xrightarrow{\text{SLOW}}$ APoA-I$_D$-DMPC*+ AI$_D$

           POLYMERIC VESICLES   DISSOCIATED   COMPLEX

N > 100   APoA-I$_P$ + DMPC $\xrightarrow{\text{FAST}}$ APoA-I$_D$ $\xrightarrow{\text{FAST}}$ APoA-I$_D$-DMPC $\xrightarrow{\text{SLOW}}$ APoA-I$_D$-DMPC*

HUMAN APOA-I + 12 MOLE OH-PC

NATIVE BABOON APOA-I

N < 100   APoA-I$_D$ + DMPC $\xrightarrow{\hspace{1cm}\text{FAST}\hspace{1cm}}$ APoA-I$_D$-DMPC $\xrightarrow{\text{SLOW}}$ APoA-I$_D$-DMPC*

           DISSOCIATED   VESICLES   COMPLEX

N > 100   APoA-I$_D$ + DMPC $\xrightarrow{\hspace{1cm}\text{FAST}\hspace{1cm}}$ APoA-I$_D$-DMPC $\xrightarrow{\text{SLOW}}$ APoA-I$_D$-DMPC*

Fig. 7C.   Kinetics of dimyristoyl lecithin vesicles
           association with native human and baboon
           apoA-I protein.

IV. MONITORING OF THE REASSEMBLY PROCESS

A. Time progress of the reaction

Fast monitoring of the reaction would involve the use of a
stopped-flow spectrophotometer or spectrofluorimeter as described
by POWNALL et al (37) for the transfer of pyrene between lipoproteins.

Monitoring over a longer time period can be carried out by various techniques : light scattering detects decrease of the light
intensity at 90°, due to the decrease of the size of the complex
and to changes in refractive index (39). By electron microscopy a
decrease of the size of the multilamellar vesicles and the formation
of stacked-discs or "rouleaux" can be visualized (40).

Microcalorimetry does not yield information about the kinetics
of fast reactions. However when slow processes are superimposed on
the fast association reaction, both effects can be analyzed separately. We applied this technique to the analysis of the apoA-I association with small vesicles of dimyristoyl lecithin (38). On the
thermograms depicted in Fig. 7A, the decay curve is slower for
apoA-I than for apoA-II-dimyristoyl lecithin (DMPC) association.
The thermogram corresponding to apoA-I-DMPC association can be resolved in a fast and slow process by comparison with the experimental curve for apoA-II (Fig. 7B). The fast exothermal process was
attributed to dissociation of polymeric apoA-I and binding to DMPC
while the slow process would arise from a size decrease and structural reorganization of the complex. This slow process is also
temperature dependent and minimal at the phospholipid transition
temperature where optimal reassembly occurs (Fig. 7C). Human apoA-I
preincubated with lysolecithin and native baboon apoA-I with a low
polymer content associate readily with the liposomes.

B. Complex isolation

After completion of the reaction period, the mixture contains
free proteins and lipids and a distribution of lipid-protein complexes. Three methods have mainly been applied to the isolation
of the various compounds.

1. Sequential ultracentrifugation

This method, developped for the prurification of lipoprotein
classes, has also been applied to the isolation of lipid-protein
complexes. Free lipids have a density less than 1.06 g/ml, free
proteins sediment around 1.23 g/ml, while the complexes have an in-
termediary density depending on their lipid to protein ratio. Sequential ultracentrifugation involves two steps. Free lipids are
first floated at the top at a density of 1.063 g/ml at 200.000 g
for 24h. The density of the infranatant is subsequently adjusted
at 1.21 g/ml and the complex isolated in the supernatant after a

24h spin at 200.000 g. The protein composition of the complex can
be subsequently determined by the method of LOWRY (41) or by amino
acid quantitation of an aliquot. The phospholipid content can be
derived from a phosphorus analysis according to either BARLETT (42)
or EIBL and LANDS (43).

A limitation of the sequential ultracentrifugation technique
is the lack of information about the complex density and the possible dissociation of lipid-protein interactions due to 48h spin-
ning in a salt medium.

ASSMANN and BREWER (44) first applied this technique to the
analysis of the complexes formed between human apoA-I and apoA-II
protein and various phospholipids. We used this method in conjunction with isothermal microcalorimetry to characterize the association of human apo-HDL and apoA-I protein with lysolecithin and dimyristoyl lecithin (45).

2. Gradient ultracentrifugation

A more informative approach to the separation of the components
of the reaction mixture is the use of density gradients formed in
the preparative ultracentrifuge. Even in this case prolonged ultracentrifugation is required and complex formation might be disturbed
by the high concentrations of the gradient components (46).
However an advantage of this technique is, that it provides a more
detailed picture of the lipid-protein species, assuming that the
analysis have been conducted after equilibrium has been reached.

We applied this technique to the isolation and characterization of the complexes formed between human and baboon apoA-I and
apoA-II-dimyristoyl lecithin and sphingomyelin. The experimental
procedure is as follows : linear gradients made of NaBr spanning
the density range 1.06-1.21g/ml were prepared by means of a gradient
mixing device and 4.6ml tubes were filled by means of a Büchler
Autodensiflo. An approximate amount of 1mg apoprotein + 3mg lipid
dispersed in the 1.06g/ml density solution were applied to each
tube. Gradient were spun for 66h at 220.000 x g in a SW50 rotor in
the Beckman L5-65 centrifuge and subsequently eluted by means of
the Autodensiflo.

The optical density at 280nm was recorded by passage through a
Pye-Unicam spectrophotometer equipped with a 8 μl cell and 0.3ml
fractions were collected. The refractive index of each fraction
was measured and the corresponding density calculated to obtain the
gradient profiles. Protein and lipids were quantitated in each
fraction, according to the methods described above.

Fig. 8A and B illustrate the patterns obtained after association of dimyristoyl lecithin and sphingomyelin with human and baboon apoA-I protein. Complex composition and yields are summarized
in Table 1. together with the data obtained on dimeric human apoA-II
and monomeric baboon apoA-II proteins (39).

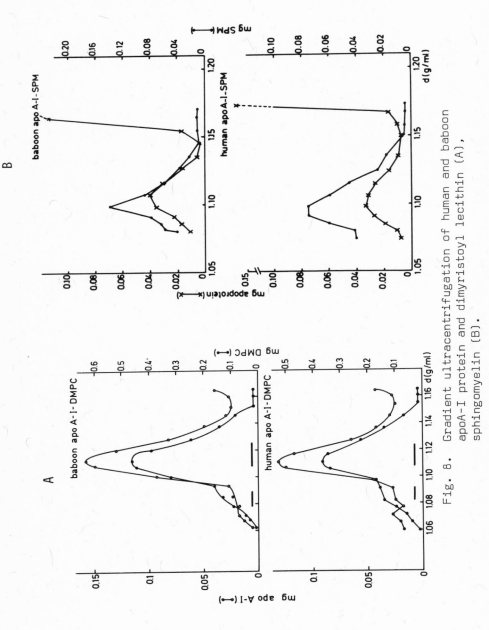

Fig. 8. Gradient ultracentrifugation of human and baboon apoA-I protein and dimyristoyl lecithin (A), sphingomyelin (B).

Table 1

APOPROTEIN	Protein + DMPC				Protein + SPM		
	d	n	a	b	d	n	b
ApoA-I (H)	1.09	$254.0^* \pm 12$	14		1.10	84.0 ± 8	49
				78.4			
	1.12	122.0 ± 2	86				
ApoA-I (B)	1.08	248.1 ± 3	20		1.10	79.0 ± 7	31
				81.5			
	1.12	126.0 ± 3	80				
ApoA-II (H)	1.10	59.8 ± 2	54		1.09	55.5 ± 5	88
				74.5			
	1.12	45.0 ± 3	46				
ApoA-II (B)	1.10	29.3 ± 1.2	73		1.09	24.3 ± 3	89
				82.1			
	1.12	22.1 ± 2.0	27				

DMPC : dimyristoyl-L-alpha-phosphatidylcholine
SPM : sphingomyelin
* : mean value of three individual determinations
d : density g/ml

n : $\dfrac{\text{Mole DMPC}}{\text{Mole apopr.}}$ or $\dfrac{\text{Mole SPM}}{\text{Mole apopr.}}$

a : amount of complex (%)
b : % apoprotein bound
H : human
B : baboon.

These results suggest that dimyristoyl lecithin associates more easily than sphingomyelin with the apoA-I protein, while the affinity of the apoA-II protein for both phospholipids is similar. The lipid-binding properties of human and baboon apoproteins are very close, as could be expected from the analogies in their amino acid sequence (30). Those data demonstrate that gradient ultracentrifugation offers advantages such as high resolution, good reproducibility and low sample amount. In case of a large lipid excess a first centrifugation step at 1.063g/ml, prior to applying the gradient, avoids any contamination by free lipids.

3. Gel chromatography

Separation of the various components according to their molecular size was performed on Sepharose 6B, 4B and on Biogel AM5 columns equilibrated either in Tris-HCl buffer pH 8.6 or in NH_4HCO_3

0.05M pH 9.0 (47, 44). We isolated the complexes formed between
sphingomyelin liposomes and human apoA-I and apoA-II proteins on a
Sepharose 6B column (Fig. 9A and B). Free sphingomyelin elute within the void volume while the complexes have a molecular weight of
300.000. 50% free apoA-I protein was still present even though an
excess of lipid was present in the reaction mixture.

    These results support the data obtained by gradient ultracen-
trifugation reported above and moreover provide information about
the molecular weight of the complex.

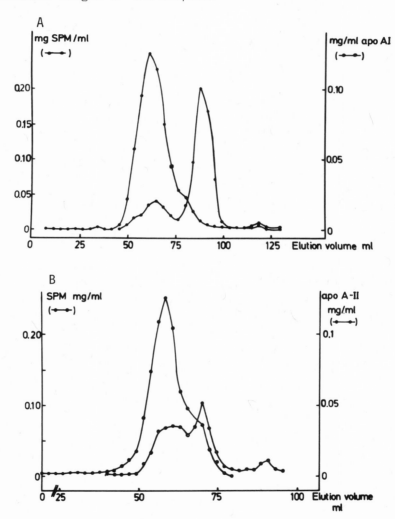

Fig. 9.   Isolation of the complexes formed between the
          sphingomyelin-apoA-I protein (A) and apoA-II
          protein (B) on Sepharose 6B, 0.05M NH_4HCO_3 pH 9.0

V. CHARACTERIZATION OF THE LIPID-PROTEIN COMPLEXES

A. Compositional analysis

The chemical analysis of the lipid and protein moiety of the reassembled lipoproteins can be performed according to the standard procedures for lipoproteins (35).
A modification of the Lowry procedure was proposed by HERBERT et al (48) to avoid turbidity due to phospholipids i.e. the addition of 0.1ml 0.5M sodium dodecyl sulfate to the reaction mixture.

B. Morphological analysis

By electron microscopy, reassembly mixtures consisting of either apoA-I, apoA-II, apoC-I or apoC-III protein and lecithin appear as stacked discs or "rouleaux" (40). This pattern is changed, by the addition of cholesterol esters, into spheres whose dimensions are similar to those of the native HDL particles (49).
Gel filtration (47), light scattering (39) and sedimentation velocity (33) analysis provide information about molecular weight and size of the reassembled complexes. From fluorescence depolarisation measurement the equivalent volume can be calculated while low-angle X-ray scattering yield the radius of the particle.

C. Structural analysis

The organization of the lipid and protein components within the complex can be derived from the combination of various physical techniques. The secondary structure of the apoprotein is calculated from circular dichroism and optical rotary measurements (50). The calculation of the most probable helical segments of the apoprotein provides information about the nature of the amino acids involved in phospholipid binding (35). The environment of the aromatic amino acids such as tyrosine or tryptophan either in the apoprotein or the complex can be analyzed by differential U.V. spectroscopy or by measurement of the intrinsic tryptophan fluorescence (51).
Potentiometric and calorimetric titration of acidic and basic amino acid residues in the apoA-I, apoA-II, ApoC-I, apoC-III protein and in their complexes with lecithin have demonstrated that a number of charged residues are masked in the complex suggesting that some ionic interactions contribute to the structural stabilization of the complex (52).
The organization of the lipid phase and more specially the degree of freedom the apolar groups can be studied by NMR. ^{13}C, ^{1}H, ^{31}P NMR have provided information about the location and the nature of the hydrophobic interactions within the complex (53). Electron spin resonance, fluorescence depolarization and differential scanning calorimetry have been applied to thermal analysis of the fluidity and phase transition of the lipids. Low angle X-ray

scattering yields the election density distribution and locate the polar and apolar components of the complex (54).

D. Thermodynamical analysis

The thermodynamics of lipid-protein interaction are complex as they result from various contributions such as conformational chan- ges of the apoprotein and the phospholipid, transfer from some re- sidues from a polar to a more apolar environment, etc. (45). Though only limited information is available, the change in free energy ΔG can be calculated from the kinetics of a lipid-protein association or from equilibrium dialysis experiments, the enthalpy change ΔH can be measured directly by microcalorimetry (52). The entropy change ΔS can be calculated from the above data and the con- tribution of ionic and hydrophobic forces evaluated.

VI. INTERACTION OF HDL APOPROTEINS WITH LIPIDS

A. ApoA-I protein

The complexes formed between HDL apoproteins and phospholipids appear as particles stacked in linear arrays with a periodicity of about 50Å when viewed by electron microscopy after negative staining. The individual subunits are discs, with a diameter ranging from 100-200Å. Addition of cholesteryl esters resulted in particles ha- ving a morphology similar to that of HDL (40, 49).

By gel filtration and electron microscopy HAUSER et al descri- bed the formation of a 200.000 dalton complex between porcine HDL and DMPC (55). We obtained a molecular weight of $2.3.10^5$ daltons for human and baboon apoA-I-DMPC complexes. This complex includes 50% of the apoA-I protein in the reaction mixture (39).

Interaction with sphingomyelin was possible only to a lesser extent (table 1). According to STOFFEL and DÄRR (56) the presence of lysolecithin was necessary in the reaction mixture to obtain high yields of complex formation with a ratio 1:48:13 for apoA-I/ lecithin/lysolecithin.

Isothermal microcalorimetry carried out at 28° indicated that apoA-I binds dimyristoyl lecithin exothermally to form a complex which incorporates 80% of the protein in the reaction mixture. The complex composition is 130mole dimyristoyl lecithin/mole apoA-I protein and the enthalpy change is -220Kcal/mole apoA-I. Part of this enthalpy change is due to infolding and increase in helicity of apoA-I under binding (45). This enthalpy change was characteri- zed by an endothermal component at low phospholipid-protein ratios which was attributed to apoA-I dissociation and unfolding and could be reduced by preincubation with lysolecithin. This effect was less pronounced in the baboon apoA-I consisting mainly of dimers instead of dimers and titrations for the human proteins (31).

This incomplete dissociation of apoA-I is an important factor influencing its recombination with phospholipids. Though it can

be overcome by reassembly with small unilemellar DMPC vesicles, it
impedes reassembly with egg-lecithin and sphingomyelin heterogenous
vesicles. This might account for the literature discrepancies concerning phospholipid binding to this apoprotein.

    We measured the microviscosity of dimyristoyl lecithin lipo-
somes after labelling with 1.3 diphenyl, 1, 3, 5 hexatriene (DPH) by
following the degree of fluorescence polarization as a function of
temperature (Fig. 10.).

    In the presence of increasing quantities of apoA-I protein the
phospholipid transition occuring at 24° disappears gradually, while
a new transition appears shifting to higher temperatures under addition of apoA-I (Table 2).

    The presence of apoA-I fluidizes the acyl chains at low tempe-
ratures and increases η at t > transition temperature.

B. ApoA-II protein

    The second major component of apo-HDL has been characterized
by several authors with respect to its sequence, self-association
properties and lipid-binding characteristics (50). This apoprotein
occurs either as a dimer in man or chimpanzee or a monomer in rhe-

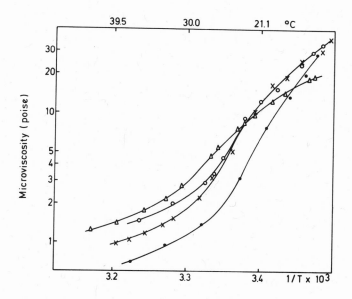

Fig. 10.   Microviscosity of dimyristoyl lecithin
           apoA-I protein mixtures   () DMPC; (x)
           n = 120; (o) n = 60; (Δ) n = 17 $\frac{\text{mole DMPC}}{\text{mole apoA-I}}$

Table 2 Phase transition parameters of DMPC/apoprotein complexes

Complexes	Microviscosity (poise) 15°C	40°C	Low Temp. Transition T_{M1}	Range	ΔT	High Temp. Transition T_{M2}	Range	ΔT
DMPC	18.00	0.63	24.0	21.9 - 26.1	4.2			
Mol DMPC Mol ApoA-I								
N=120	24.29	0.91	23.5	21.8 - 25.2	3.4	25.5	23.7 - 28.4	4.7
N= 60	24.29	1.13				25.5	23.6 - 28.3	4.7
N= 17	15.04	1.29				27.5	25.6 - 33.0	7.4
Mol DMPC Mol ApoA-II								
N= 70	26.75	1.08	24.0	21.9 - 27.3	5.4	27.5	23.3 - 31.4	8.1
N= 35	22.21	1.26				28.0	24.0 - 32.4	8.4
N= 16	10.43	1.38				29.5	24.8 - 32.4	7.6
Mol DMPC Mol ApoC-I								
N=125	25.06	0.79	23.5	21.0 - 26.0	5.0	25.5	22.5 - 29.0	6.5
N= 62	26.00	0.89				25.5	21.8 - 29.2	7.4
N= 21	25.06	1.12				26.5	22.7 - 30.2	7.5
Mol DMPC Mol ApoC-III								
N=120	32.80	0.70	24.0	21.8 - 26.2	4.4	26.5	24.2 - 31.2	7.0
N= 65	26.75	0.74	24.0	21.9 - 26.1	4.2	27.0	22.8 - 31.2	8.4
N= 21	32.07	0.84	26.0	22.7 - 29.6	6.9	28.5	25.4 - 31.6	6.2

sus and baboon. The presence of the cysteine does not affect its affinity for phospholipids (30). According to LUX et al (1), ASSMANN and BREWER (44), the lipid-binding segments in human apoA-II protein are located at the carboxyl-terminal end of the protein. Our potentiometric data support this interpretation and suggest that 6 out of 9 lysines are masked by phospholipids in the apoA-II lecithin complex (57, 58).

We have compared the affinity of human and baboon apoA-II protein for lecithin and sphingomyelin and obtained similar data for both phospholipids, whereas for apoA-I protein a preferential binding to lecithin was observed (Table 1). Moreover the kinetics of the apoA-II-phospholipid association are faster than for apoA-I protein and the reaction more exothermal (Fig. 7A). This apoprotein seems therefore to bind lipids more readily than apoA-I protein. The shift of the transition temperature of dimyristoyl lecithin is more significant upon addition of apoA-II than apoA-I protein. The same was observed for the fluidization of the acyl chains at t >

transition temperature and their immobilization at t $<$transition temperature (Fig. 11.).

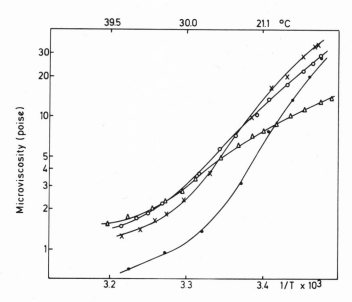

Fig. 11. Microviscosity of DMPC-apoA-II mixtures
(•) DMPC; (x) n = 70; (o) n = 35; (Δ)n=16

$$n = \frac{mole\ DMPC}{mole\ apoA\text{-}II\ protein}$$

VII. INTERACTION OF VLDL APOPROTEINS WITH LIPIDS

The interaction of apoC-I, apoC-II, apoC-III proteins with lipids is well documented and presents a number of characteristics in
common (59, 60).

An increase in helical content as well as a shift in the tryptophan fluorescence maximum was observed, the complex composition
was 60 moles lecithin/mole apoC-III protein. We measured the enthalpy change on binding dimyristoyl lecithin to apoC-I and apoC-III
protein (61) and observed the changes in microviscosity during complex formation.

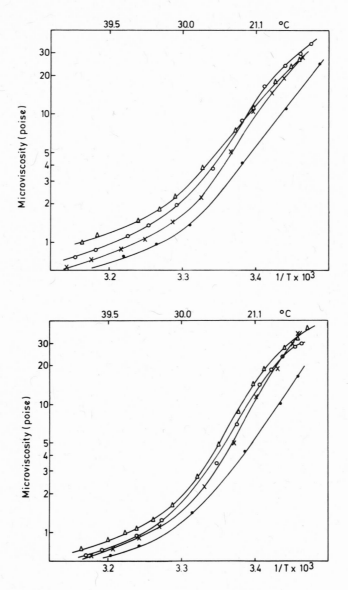

Fig. 12. Microviscosity of apoC-I/DMPC (A) and
apoC-III/DMPC (B) mixtures
A (•) DMPC; (x) n = 125; (o) n = 62;
(Δ) n = 21mole DMPC/apoC-I

B (•) DMPC; (x) n = 120; (o) n =65;
(Δ) n = 21mole DMPC/mole apoC-III

By comparison of the titration data of the native apoproteins and of their complexes with lecithin, location and structure of lipid-binding segments was proposed. These results agree with those of relipidation studies on native or synthetic apoC-I and apoC-III fragments (58).

CONCLUSIONS

The reassembly of apoproteins and lipids involves the knowledge of the nature of the constituents, of their physical properties in solution and the analysis of their behaviour in the reassembled state. Thanks to the use of model systems, the information in this area becomes more accurate and more relevant to the physiological function. An ultimate test on the validity of any "in vitro" reassembly experiment is of course the analysis of the biological activity of the system and its aptitude to simulate the "in vivo" situation. Through this bias the use of such systems either semi or completely synthetic might prove valuable to replace deficient functioning entities for lipid transport and turnover during genesis and development of atherosclerosis.

ACKNOWLEDGEMENTS

We are endebted to Mrs.J.LIEVENS, Mr.R.VERCAEMST and Mr.H.CASTER for skilful technical assistance. This research was supported by Grant 3001075 NFGWO, Brussels.

REFERENCES

1. S.E.LUX, R.HIRZ, R.I.SHRAGER, A.M.GOTTO
    J.Biol.Chem. _247_, 2598 (1972).

2. A.M.SCANU, C.T.LIM, C.T.EDELSTEIN
    J.Biol.Chem. _247_, 5850 (1972).

3. A.M.SCANU, E.CUMP, J.TOTH, S.KOGA, E.STILLER, J.ALBERS
    Biochemistry _9_, 1327 (1970).

4. R.B.HAMILTON, H.J. KAYDEN
    "Biochemistry of Disease", 5, 531, F.F.Becker, ed., Dekker, New York (1974).

5. C.TANFORD
    "The Hydrophobic Effect" p.121
    Wiley, New York (1973).

6. R.B.VERDERY, A.NICHOLS
    Chem.Phys.Lipids _14_, 123 (1975).

7.   M.ROSSENEU, F.SOETEWEY, V.BLATON, J.LIEVENS, H.PEETERS
     Chem.Phys.Lipids 13, 203 (1974).

8. H.HAUSER, M.C.PHILLIPS
     J.Biol.Chem. 248, 8585 (1973).

9.   A.N.KRUSKI, A.M.SCANU
     Chem.Phys.Lipids 13, 27 (1974).

10. V.LUZZATI
     in "Biological Membranes" p.71
     ed. D.Chapman, Acad.Press, New York (1968).

11. R.M.WILLIAMS, D.CHAPMAN
     Prog. Chem. Fats and other Lipids 11, 1 (1970).

12. D.CHAPMAN, D.F.H.WALLACH
     in "Biological Membranes" p.125,
     D.Chapman ed., Acad.Press, New York (1968).

13. A.D.BANGHAM, R.W.HORNE
     J.Mol.Biology 8, 1660 (1964).

14. K.J.PALMER, F.O.SCHMITT
     J.Cell.Comp.Physiol. 17, 385 (1941).

15. A.D.BANGHAM, M.M.STANDISH, J.C.WATKINS
     J.Mol.Biol. 3, 238 (1965).

16. C.H.HUANG
     Biochemistry 8, 344 (1969).

17. J.SURKUUSK, B.R.LENTZ, Y.BARENHOLZ, R.L.BILTONEN, T.E.THOMPSON
     Biochemistry 15, 1393 (1976).

18. M.C.PHILLIPS, H.HAUSER, R.B.LESLIE, D.OLDANI
     Biochim.Biophys.Acta 406, 402 (1975).

19. S.BATZRI, E.D.KORN
     Biochim.Biophys.Acta 298, 1015 (1973).

20. J.BRUNNER, P.SKRABAL, H.HAUSER
     Biochim.Biophys.Acta 455, 322 (1976).

21. B.LUNDBERG
     Chem.Phys.Lipids 18, 212 (1977).

22. B.LUNDBERG, E.R.SAARINEN
     Chem.Phys.Lipids 14, 260 (1975).

23. B.SHORE, V.SHORE
     Biochemistry 8, 4510 (1969).

24. W.V.BROWN, R.I.LEVY, D.S.FREDRIKSON
     J.Biol.Chem. 245, 6588 (1970).

25. V.BLATON, R.VERCAEMST, N.VANDECASTEELE, H.CASTER, H.PEETERS
     Biochemistry 13, 1127 (1974).

26. J.C.OSBORNE, G.PALUMBO, B.BREWER
    Biochemistry 15, 317 (1976).

27. J.GWYNNE, H.B.BREWER, H.EDELHOCH
    J.Biol.Chem. 250, 2269 (1975).

28. L.B.VITELLO, A.M.SCANU
    J.Biol.Chem. 251, 1131 (1976).

29. L.B.VITELLO, A.M.SCANU
    Biochemistry 15, 1161 (1976).

30. V.BLATON, R.VERCAEMST, M.ROSSENEU, J.MORTELMANS, R.L.JACKSON,
    A.M.GOTTO, H.PEETERS
    Biochemistry 16, 2157 (1977).

31. M.ROSSENEU, V.BLATON, R.VERCAEMST, F.SOETEWEY, H.PEETERS
    Eur.J.Biochem. 74, 83 (1977).

32. A.JONAS
    Biochim.Biophys.Acta 393, 471 (1975).

33. W.L.STONES, J.A.REYNOLDS
    J.Biol.Chem. 250, 8045 (1975).

34. A.M.SCANU
    J.Lipid.Res. 7, 295 (1966).

35. A.M.SCANU, C.EDELSTEIN, P.KEIM
    in "The Plasma Proteins : Structure, Function and Genetic
    Control",
    F.Putnam ed. Acad.Press, New York, p.317 (1975).

36. M.ROSSENEU, F.SOETEWEY, R.VERCAEMST, J.LIEVENS, H.PEETERS
    Prot.Biol.Fluids 25, 47 (1977).

37. H.J.POWNALL, J.B.MASSEY, J.T.SPARROW, A.M.GOTTO
    Prot.Biol.Fluids 25, 115 (1977).

38. M.ROSSENEU, F.SOETEWEY, J.LIEVENS, H.PEETERS
    Prot.Biol.Fluids 25, 121 (1977).

39. M.ROSSENEU
    Aggregaatsthesis, Univ. of Antwerpen (1977).

40. T.M.FORTE, A.V.NICHOLS, E.L.GONG, S.LUX, R.I.LEVY
    Biochim.Biophys.Acta 248, 381 (1971).

41. O.H.LOWRY, H.J.ROSEBROUGH, A.L.FARR, R.J.RANDALL
    J.Biol.Chem. 193, 265 (1951).

42. G.R.BARTLETT
    J.Biol.Chem. 234, 466 (1959).

43. H.EIBL, W.E.LANDS
    Biochemistry 9, 423 (1970).

44. G.ASSMAN, H.B.BREWER
    Proc.Nat.Acad.Sci.US, 47, 1309 (1974).

45. M.ROSSENEU, F.SOETEWEY, G.MIDDELHOFF, H.PEETERS, W.V.BROWN
    Biochim.Biophys.Acta 441, 68 (1976).

46. A.M.SCANU, J.L.GRANDA
    Biochemistry, 5, 446 (1966).

47. G.MIDDELHOFF, M.ROSSENEU, H.PEETERS, W.V.BROWN
    Biochim.Biophys.Acta 441, 57 (1976).

48. P.N.HERBERT
    Personal Communication.

49. T.FORTE, E.GONG, A.V.NICHOLS
    Biochim.Biophys.Acta 551, 169 (1974).

50. J.D.MORRISETT, R.L.JACKSON, A.M.GOTTO
    Ann.Rev.Biochem. 44, 183 (1975).

51. C.J.HART, R.B.LESLIE
    Chem.Phys.Lipids 4, 369 (1970).

52. H.PEETERS, M.ROSSENEU
    in "Living Systems as Energy Converters"
    R.Buvet, J.P.Massué ed., Elsevier Publ. p.55(1977).

53. B.SEARS, R.J.DECKELBAUM, M.J.JANIAK, G.G.SHIPLEY, D.M.SMALL
    Biochemistry 15, 4151 (1976).

54. G.G.SHIPLEY, D.ATKINSON, A.M.SCANU
    J.Supramol.Struct. 1, 98 (1972).

55. H.HAUSER, R.HENRY, R.B.LESLIE, J.M.STUBBS
    Eur.J.Biochem. 48, 583 (1974).

56. W.STOFFEL, W.DÄRR
    Hoppe-Zeyler's Z. Physiol.Chem. 357, 127 (1976).

57. M.ROSSENEU, F.SOETEWEY, M.J.LIEVENS, R.VERCAEMST, H.PEETERS
    Eur.J.Biochem. 79, 251 (1977).

58. F.SOETEWEY, M.J.LIEVENS, R.VERCAEMST, M.ROSSENEU, H.PEETERS,
    V.W. BROWN
    Eur.J.Biochem., 79, 259 (1977).

59. R.L.JACKSON, J.D.MORRISETT, R.T.SPARROW, J.P.SEGREST,
    H.J.POWNALL, L.C.SMITH, H.F.HOFF, A.M.GOTTO
    J.Biol.Chem. 249, 5314 (1974).

60. J.D.MORRISETT, J.S.K.DAVID, H.J.POWNALL, A.M.GOTTO
    Biochemistry 12, 1290 (1973)

61. M.ROSSENEU, F.SOETEWEY, H.PEETERS, L.BAUSSERMAN, P.N.HERBERT
    Eur.J.Biochem., 70, 285 (1976).

Metabolism of the Lipoprotein Molecule

INTERCONVERSION OF PLASMA LIPOPROTEINS

Alex V. Nichols

Donner Laboratory, Lawrence Berkeley Laboratory

University of California, Berkeley, CA. 94720 USA

INTRODUCTION

Interconversion of a lipoprotein molecule to a lipoprotein molecule of different physical and chemical properties perhaps can be best considered in terms of our current understanding of lipoprotein structure. Thus, normal lipoproteins are generally described as spherical structures with a core of triglycerides and cholesteryl esters stabilized by a surface of specific apoproteins, phospholipids and cholesterol (1, 2). On the surface, the phospholipids and cholesterol form a monolayer in which apoproteins exhibiting amphipathic helical regions are embedded. This overview will consider interconversions which result from significant alterations in molecular size and/or shape due to enzymatically or physico-chemically-induced changes in core and/or surface domains of a lipoprotein molecule. Analytic approaches to the investigation of interconversions include: preparative and analytic ultracentrifugation, electron microscopy, gel chromatography and electrophoresis, and chemical analysis.

INTERCONVERIONS: ENZYME-INDUCED

Lipoprotein Lipase: Interconversion of Very Low Density Lipoproteins

Hydrolytic depletion of core triglycerides in very low density lipoproteins (VLDL) by post-heparin lipolytic activity (PHLA) produces lipoprotein species of smaller size and higher particle density (3). The chemical properties of these species indicate a process of core reduction accompanied by decrease of surface material in excess of that required to stabilize the remaining triglyceride core.   Transfer of C apoproteins and phospholipid from the VLDL surface to high density lipoproteins (HDL) during removal of VLDL triglyceride from the circulation has been described (4) and would appear to be associated with the process of surface adjustment during hydrolytic core reduction.   In certain states of abnormal lipid metabolism, accumulations of lipoprotein surface lipids are observed in vivo in the form of abnormal particles in the plasma.   Thus, in patients deficient in plasma lecithin:cholesterol acyltransferase (LCAT), abnormal vesicular species rich in phospholipid and cholesterol appear in the d 1.006-1.019 and d 1.019-1.063 fractions and are considered to arise during degradation of chylomicrons and VLDL by lipoprotein lipase (LPL) in vivo (6).   Such vesicular species have not been observed in plasma of normal subjects and it would appear that disposal of surface components is efficiently handled under normal conditions.   Enzymatic mechanisms proposed for disposition of lipoprotein surface lipids include: (1) hydrolysis of the phospholipids by phospholipase activity associated with LPL, and (2) conversion by LCAT of surface phosphatidylcholine and unesterified cholesterol to lysolecithin and cholesteryl esters (7).   The interconversions noted in triglyceride-rich lipoproteins during PHLA activity are considered similar to those occurring during normal metabolic clearance of these species, where "remnants" of chylomicrons and VLDL are formed during hydrolytic encounters with LPL situated on capillary endothelial surfaces (8).   An interesting aspect of the above interconversion pathway is the apparent persistence of association of one major apoprotein (apoB) with the residual core lipid leading to the eventual formation of the low density lipoprotein (LDL) class (9).   Differential affinities of other apoproteins for the lipid moiety remaining during the course of VLDL interconversion are also suggested by the differing apoprotein compositions of VLDL subspecies of increasing hydrated density (10). One of the released apoproteins is the activating cofactor for LPL (apoC-II).   Precursors to VLDL particles secreted by the liver are also spherical particles but of somewhat different chemical composition particularly at the surface (11, 12).   These particles pick up C apoproteins from HDL in blood and with some additional transfers of surface lipid assume the composition of VLDL usually encountered in blood (13).   Apparently, no other major structural transformations of precursor plasma chylomicrons and VLDL occur until the particles encounter the hydrolytic activity of LPL.

Lecithin:Cholesterol Acyltransferase:Interconversion
of Discoidal to Spherical HDL

Another major interconversion process occurring in vitro and probably in vivo involves the enzymatic build-up rather than degradation of an apolar lipid core in a lipoprotein molecule.   This interconversion has been designated the "disc-to-sphere" interconversion wherein discoidal lipoprotein complexes, composed primarily of phospholipid, unesterified cholesterol and specific apoproteins and deficient in core lipids, are transformed by LCAT to spherical structures during build-up of an apolar core of cholesteryl ester (14, 15, 16). The spherical particles formed have properties similar to normal HDL. Discoidal complexes detected in the d 1.063-1.20 fraction of plasma of human subjects with LCAT-deficiency and cholestasis transform to spherical HDL particles when plasma of such subjects is incubated with LCAT (14). The similarity of the spherical products to normal HDL and the occurrence of discoidal particles in plasma of certain human patients have prompted the proposal that such discoidal complexes may be precursor HDL which are converted to normal HDL after secretion and interaction with LCAT.

Rat liver perfusion studies. Animal data supporting the idea that discoidal complexes are precursor forms of HDL (nascent HDL), secreted by the liver and converted to spherical HDL by LCAT activity, have recently been reported by Hamilton et al. (17). Using the perfused rat liver, these workers demonstrated discoidal complexes (diameter: 190 ± 25 Å; thickness: 46 ± 5 Å) in a perfusate containing an LCAT inhibitor (5,5′-dithionitrobenzoic acid, DTNB). In the presence of DTNB the perfusate contained predominantly (∿ 75% of particles observed) spherical species (diameter: 75-125 Å) which were similar to normal rat HDL (diameter: 114 ± 13 Å). The discoidal complexes were rich in surface lipids (phospholipids and unesterified cholesterol) and deficient in core lipids. The HDL fraction from the noninhibited perfusate (containing spherical particles and some discs) showed a level of core lipid approaching, but not equal, to that of normal rat HDL. The apoprotein composition by polyacrylamide gel electrophoresis was markedly different in the discoidal and spherical HDL preparations. Hence, while apoE (arginine-rich apoprotein) was found to be the major component of the discoidal species from the inhibited perfusate, apoA-I was the primary component of the HDL from the noninhibited perfusate rich in spherical particles. The above would suggest that during the LCAT-induced disc-to-sphere interconversion, apoA-I is incorporated and apoE is to some extent dissociated. However, an alternative explanation might be that ultracentrifugation in high salt dissociates apoA-I from discoidal complexes and apoE from spherical HDL. Indeed, results consistent with dissociation of some apoE from rat plasma spherical HDL (rich in apoA-I) during ultracentrifugation have been reported (18, 19). Although dissociation of apoA-I from rat discoidal complexes during ultracentrifugation has not been evaluated, possible dissociation of apoA-I from lipoproteins rich in surface lipids (i.e., discs) has been noted during ultracentrifugation of LCAT-deficient plasma (20). Thus, following preparative ultracentrifugation of LCAT-deficient plasma, there is considerable apoA-I in the d > 1.25 fraction. Interestingly, a substantial amount of this apoA-I is incorporated into the d 1.063-1.20 fraction following incubation of the plasma with LCAT, suggesting that under conditions of ultracentrifugation, apoA-I may have a greater affinity for cholesteryl ester-rich particles than for surface lipid-rich particles (20). Hence, in terms of the available data, it is not possible to establish whether the observed apoprotein compositional differences reflect an interconversion process whereby apoA-I (from some unidentified source) is incorporated and some apoE is dissociated during formation of a cholesteryl ester core in discoidal particles; or whether both apoproteins are initially present in the discs and spheres but are dissociated during ultracentrifugation in accord with their differential affinities for lipid mixtures deficient or enriched in core lipid. At present no data are available on the physical-chemical changes occurring in discoidal complexes during in vitro incubation with LCAT of a rat liver perfusate collected under conditions where LCAT was inhibited. Hence, no comparison is possible between spherical products produced utilizing perfusate discoidal complexes as substrate and those appearing in plasma or in the noninhibited perfusate of rat liver where the conversion is presumed to occur upon secretion of discoidal complexes into an LCAT-containing medium. Clearly the sequence of events whereby the final apoprotein composition of normal rat plasma HDL is attained, during the disc-to-sphere interconversion, requires further investigation.

Additional definition of the molecular species involved in the interconversion process is probably also needed. For example, electrophoresis of the HDL from the rat liver perfusate, in which LCAT activity is inhibited, indicates the presence of two lipoprotein species, whereas predominantly one (discoidal complexes) is observed under electron microscopy (17). Data from studies on LCAT-deficient plasma have demonstrated the participation of HDL species (small molecular weight HDL particles with diameters of 50-60 Å) other than discoidal complexes in the interconversion process (21). Thus, the interconversion process may involve not only direct transformation of discs to spheres, but also participation of some intermediate subspecies which could act as alternate sites for cholesteryl ester build-up and formation of particles similar to normal HDL.

Lastly, in view of the apparent ability of LCAT to transform diverse substrate particles to relatively homogeneous HDL species, it is possible that the precise form of the substrate is not crucial and that an LCAT-apoA-I complex is the critical HDL-generating unit which functions by organizing apoprotein and polar lipid moieties around a growing cholesteryl ester core. Inherent lipid binding affinities and capacities of specific apoproteins would determine the final lipoprotein composition. The above discussion indicates the need for detailed studies on purified systems where interconversions of well-defined model substrates can be followed and characterized by physical and chemical means.

Incubation of plasma of LCAT-deficient subjects with LCAT.
The properties of the HDL particle distribution in plasma of LCAT-
deficient patients, which includes discoidal and spherical small
molecular weight HDL particles, have been rather extensively characterized. Some major differences exist between HDL of LCAT-in-
hibited rat liver perfusate and HDL of LCAT-deficiency. Unlike
the discoidal HDL in the rat liver perfusate, the discoidal HDL in
LCAT-deficient plasma has a significant content of both apoA-I and
apoA-II (22). Interestingly, however, the disc-containing fractions from both sources have a substantial content of apoE (20, 23).
The small molecular weight HDL particles in LCAT-deficient plasma
also contain substantial apoA-I and apoA-II. As noted previously,
apoA-I is found not only in the d 1.063-1.20 fraction, but also in
the d > 1.25 bottom fraction. This suggests that some apoA-I either
exists freely in such plasma or is loosely bound to phospholipid-
cholesterol-rich lipoprotein species and is dissociated upon ultracentrifugation in high salt. As previously noted, in the presence
of LCAT activity, some of this apoA-I is incorporated into d 1.063-
1.20 particles resembling HDL. It would appear that formation of
apolar core material stabilizes the binding of apoA-I to lipid,
perhaps by promoting formation of additional amphipathic helical
conformation in the apoprotein. It is also of interest to note
that, during incubation of LCAT-deficient plasma with LCAT, the
content of apoE is markedly reduced in the HDL fraction, with the
released apoE now appearing primarily in the VLDL fraction (20).
A detailed investigation of the disc-to-sphere interconversion occurring during incubation of LCAT-deficient plasma with partially-
purified LCAT has been reported (20). In this study, the interconversion was followed at two levels of added LCAT activity. At a
low level of LCAT activity due to presence of p-chloromercuriphenyl-
sulfonate [PCMPS] (LCAT activity approximately one-tenth of that in
presence of mercaptoethanol [ME]), there was a partial reduction in
content of discoidal species (diameter: 170-240 Å) which form the
large molecular weight end of their HDL distribution. This reduction was associated with the appearance of spherical HDL species
(mean diameter: ∿ 78 Å) smaller than the discs but somewhat larger
than the small molecular weight HDL (diameter: 50-60 Å) usually encountered in LCAT-deficient plasma. At full LCAT activity (in presence of ME) there was further reduction in content of discoidal
complexes (almost full disappearance by electron microscopy); and a
corresponding increase in smaller molecular weight species which
now exhibited a still larger average molecular weight and size
(mean diameter: ∿ 98 Å) than either the usual small molecular weight
species or the small molecular weight product observed at low level
LCAT activity. These data suggest that LCAT, utilizing a most heterogeneous mixture of possible substrate material in LCAT-deficient
plasma (discs, small molecular weight HDL, loosely-bound apoA-I,
etc.), initially forms small spherical molecular species (diameter
∿ 78 Å) which, with further LCAT activity, are transformed to still
larger spherical species (diameter ∿ 98 Å) within the HDL par-

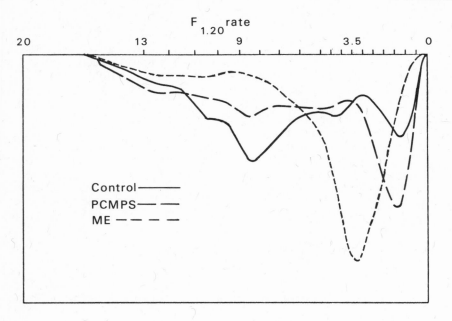

Figure 1: Analytic ultracentrifugal patterns showing the effect of incubation (24 hr, 37°C) with partially purified LCAT on the HDL in plasma of LCAT-deficient patient (M.R.) in presence of PCMPS (2 mM) and ME (10 mM).

ticle distribution. This interconversion is perhaps best appreciated by reference to the analytic ultracentrifugal patterns obtained for the above incubation mixtures (Fig. 1). In these patterns the discoidal species exhibit a peak whose flotation rate is $F^{\circ}_{1.20}$ 8.4, while the peak of the small molecular weight spherical particles exhibits a flotation rate of $F^{\circ}_{1.20}$ 1.25. The control pattern shows the presence of both of the above species. At low level LCAT activity, the build-up of smaller molecular weight species shows a peak at $F^{\circ}_{1.20}$ 1.4, while at full LCAT activity the further build-up shows a peak primarily at $F^{\circ}_{1.20}$ 3.3.

In view of the rather close correspondence of the ultracentrifugal and chemical properties of the above products to those of recently identified major components (24) within the normal HDL distribution, these observations imply that such interconversion processes may be involved in the origin of the normal HDL distribution. Thus, recent work by Anderson et al. (24) has established the existence of at least three major components within the HDL particle distribution from normal human plasma. These components have the following hydrated densities and mean peak flotation rates: HDL_{2b}, d 1.063-1.100, $F^o_{1.20}$ 5.37; HDL_{2a}, d 1.100-1.125, $F_{1.20}$ 3.15; and HDL_3, d 1.125-1.200, $F^o_{1.20}$ 1.56. Comparison of the flotation rates and chemical compositions of these three components with those of the two small molecular incubation products (obtained at low and full level LCAT activity) shows rough similarity between the low level activity product and HDL_3, and the full activity product and HDL_{2a}. Complete comparison, however, will require information on the apoprotein compositions of both the incubation products and the HDL components. Nevertheless, the above incubation data do suggest a possible sequence in the origin of the HDL particle distribution, wherein LCAT activity first organizes the known substrate and cofactor components into species resembling HDL_3, and, with the availability of sufficient substrate components and further activity, into additional species resembling HDL_{2a}.

In any consideration of such a step-wise build-up of the HDL distribution it is of interest to note the steady state distribution of the plasma total HDL concentration among the three HDL components in a population sample of normal human subjects. The results of such a three component analysis show that the distribution of total HDL concentration among the three components is significantly correlated with the total HDL concentration (25). Thus, in cases with total HDL concentrations in the range of 100 ± 5 mg/dl, over 90% of the total HDL concentration is represented by HDL_3. At higher total HDL levels (100-200 mg/dl), the increase in concentration is contributed primarily by the HDL_{2a} component with little change in HDL_3. At still higher concentrations of total HDL (200-475 mg/dl), there is further increase in HDL_{2a} levels accompanied now by the appearance and build-up of HDL_{2b} levels. At very high total HDL concentrations (> 475 mg/dl), levels of HDL_{2a} and HDL_{2b} are further increased in conjunction with the appearance of additional faster floating HDL components. Thus, the average trends in distribution of the total HDL distribution among the three components suggest processes, possibly involving LCAT, which distribute increasing plasma levels of HDL lipid substrates and apoproteins in a directed manner into lipoprotein species of larger particle size and lower hydrated density. It will be of considerable interest to establish whether such interrelationships between plasma levels of these three lipoprotein components hold in individuals during short-term reduction or elevation of their total HDL levels. Although the above considerations are for the most part speculative, they do point to some exciting and challenging problems in the structure and function of the HDL particle distribution.

INTERCONVERSIONS: RECENT PHYSICAL-CHEMICAL STUDIES

As noted earlier, lipoprotein interconversions may result from alterations either in the surface or the core domains of a lipoprotein molecule. Of considerable interest are the effects of alterations restricted predominantly to one domain, which are not complicated by concomitant major alterations in the other. Recently, two very dissimilar approaches have been applied to the HDL molecule, which dissociate some of its surface apoprotein and produce interconversions of physical-chemical interest and of potential biological significance. The first approach utilizes the protein denaturant, guanidine-HCl (GuHCl) (26) while the second uses a nonphysiologic phospholipid, dimyristoylphosphatidylcholine (DMPC) (27, 28).

Interconversions of HDL Induced by Guanidine-HCl

Perturbation of HDL structure by a sequence of GuHCl exposure and removal is characterized by: (1) selective dissociation of lipid-poor apoA-I, which follows a time-course similar to denaturation-related changes in spectroscopic parameters (e.g., negative ellipticity), (2) progressive formation of lipoprotein fusion products, which are relatively depleted in apoA-I and enriched in apoA-II, apoC and lipid, and (3) apparent reversible reassociation, following exposure of HDL to high levels of GuHCl, of both apoA-II and apoC with the lipid moiety of the denatured HDL. The progressive dissociation of apoprotein from HDL produces (1) an increase in mean peak flotation rate and (2) a reduction in area of the schlieren pattern obtained upon analytic ultracentrifugation of HDL previously exposed to GuHCl. Electron microscopy indicates a progressive increase in size of the fusion products with increasing duration of exposure of the molecules to the denaturant. Perturbation of HDL structure by directly ultracentrifuging HDL in GuHCl is characterized by: (1) selective dissociation of apoA-I in the range of 2M-3M GuHCl, (2) dissociation of apoA-I, and partially of apoA-II and apoC at 4M GuHCl, and (3) extensive dissociation of apoA-I, apoA-II, and apoC at 5M GuHCl. These studies indicate differential properties of HDL apoproteins in stabilizing HDL structure and indicate a mode of lipoprotein interconversion which results from apoprotein dissociation coupled with fusion of the residual lipoprotein constituents. This interconversion process emphasizes the important role of specific apoproteins in stabilizing the lipoprotein molecule and identifies fusion as a process which can stabilize the apolar cores of lipoproteins partially depleted of their surface components.

Interconversions of HDL Induced by Dimyristoylphosphatidylcholine

Interaction of HDL with DMPC suspensions or vesicles removes some apoprotein from the lipoproteins to form discoidal complexes comparable in morphology to those described earlier in this report. The apoprotein forming the discoidal complexes appears by polyacrylamide gel electrophoresis to be primarily apoA-I. Tall and Small

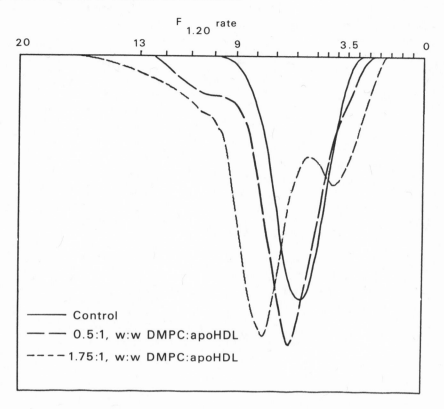

Figure 2: Analytic ultracentrifugal patterns showing the effect of interaction of HDL$_{2b}$ with DMPC vesicles.

(27) propose that the residual lipoprotein particles, which have lost some apoA-I, fuse to form particles of increased size and decreased density. Thus, this interconversion process would be similar to that put forth to explain the changes in HDL properties after exposure of HDL to GuHCl, and comparable analytic ultracentrifugal and electron microscopic information would be expected for the fusion products resulting from interaction of HDL with DMPC. Analytic ultracentrifugation (Fig. 2) of an incubated mixture of HDL$_{2b}$ (d 1.063-1.100) plus DMPC (1.75/1, w/w, DMPC/apoHDL; 37°C, 10 min., then cooled to transition temperature of DMPC) shows the transformation of the initial single HDL$_{2b}$ peak into a faster main peak and a slower minor peak (28). Separate studies have established that the slow minor peak consists primarily of discoidal complexes of DMPC and apoA-I. The area of the major peak, representing the residual lipoprotein moieties after apoprotein dissociation,

does not show a decrease in apparent concentration but an increase, suggesting uptake of DMPC either by fusion products or by HDL_{2b} which has lost some apoprotein. Utilizing radioactive DMPC we have demonstrated uptake of DMPC by the residual lipoprotein moieties after interaction of HDL_{2b} with labeled DMPC vesicles. Electron microscopy of these lipoprotein species shows spherical particles which are highly deformable upon packing under the negative stain- ing procedure used. The increased deformability is compatible with loss of surface apoprotein stabilizing the lipoprotein struc- ture. Although the dimensions of these lipoprotein particles are somewhat larger than those of the reactant HDL_{2b}, they are signi- ficantly smaller than dimensions calculated for fusion products (e.g., dimers, etc., which have lost some surface apoprotein). An estimate of the average apoprotein loss per HDL_{2b} molecule, under the experimental conditions used, yields approximately 1 apoA-I molecule removed per HDL_{2b} molecule.

At the present time, DMPC is the only phospholipid reported to effect the above interconversion. We have observed uptake of egg yolk phosphatidylcholine by HDL species but without dissocia- tion of surface apoprotein or discoidal complex formation. If under certain physiologic conditions long chain phospholipids might produce this interconversion in HDL, then this transformation could be of considerable implication to HDL metabolism as well as to interaction of HDL with membraneous structures.

ACKNOWLEDGEMENTS

This work was supported in part by National Institutes of Health grant HL-18574 and the U.S. Energy Research and Develop- ment Administration. The many valuable contributions of Drs. Trudy Forte and David Anderson and Ms. Elaine Gong, Pat Blanche and Dorothy Sprague are gratefully acknowledged.

REFERENCES

1. Shen, B. W., Scanu, A. M. and Kezdy, F. J. Proc. Natl. Acad. Sci. (USA) 74: 837, 1977.
2. Verdery, R. B. III and Nichols, A. V. Chem. Phys. Lipids 14: 123, 1975.
3. Nichols, A. V., Strisower, E. J., Lindgren, F. T., Adamson, G. L. and Coggiola, E. L. Clin. Chim. Acta 20: 277, 1968.
4. Eisenberg, S., Bilheimer, D. W., Levy, R. I. and Lindgren, F.T. Biochim. Biophys. Acta 326: 361, 1973.
5. Glomset, J. A., Norum, K. R., Nichols, A. V., King, W. C., Mitchell, C. D., Applegate, K. R., Gong, E. L. and Gjone, E. Scand. J. Clin. Lab. Invest. 35 Suppl. 142: 3, 1975.
6. Eisenberg, S. and Schurr, D. J. Lipid Res. 17: 578, 1976.

7.   Schumaker, V. N. and Adams, G. H. J. Theor. Biol. 26: 89, 1970.
8.   Olivecrona, T., Bengtsson, G., Marklund, S., Lindahl, U. and
     Höök, M. Fed. Proc. 36: 60, 1977.
9.   Eisenberg, S. and Levy, R. I. Advan. Lipid Res. 13: 1, 1975.
10. Kane, J. P., Sata, T., Hamilton, R. L. and Havel, R. J. J.
     Clin. Invest. 56: 1622, 1975.
11. Mahley, R. W., Hamilton, R. L. and LeQuire, V. S. J. Lipid
     Res. 10: 433, 1969.
12. Hamilton, R. L. and Kayden, H. J. In: The Liver: Normal and
     Abnormal Functions, F. F. Becker, ed., Marcel Dekker, Inc.,
     New York, Part A, 1974, p. 531.
13. Rubinstein, B. and Rubinstein, D. J. Lipid Res. 13: 317, 1972.
14. Forte, T., Nichols, A., Glomset, J. and Norum, K. Scand. J.
     Clin. Lab. Invest. 33 Suppl. 137: 121, 1974.
15. Hamilton, R. L., Williams, M. C., Havel, R. J. and Fielding,
     C. J. In: Atherosclerosis III, G. Schettler and A. Weizel,
     eds., Springer-Verlag, Berlin, abstr. 936, 1974.
16. Segrest, J. P. Chem. Phys. Lipids 18: 7, 1977.
17. Hamilton, R. L., Williams, M. C., Fielding, C. J. and Havel,
     R. J. J. Clin. Invest. 58: 667, 1976.
18. Fainaru, M., Havel, R. J. and Imaizumi, K. Biochim. Biophys.
     Acta 490: 144, 1977.
19. Mahley, R. W. and Holcombe, K. S. J. Lipid Res. 18: 314, 1977.
20. Norum, K. R., Glomset, J. A., Nichols, A. V., Forte, T., Albers,
     J. J., King, W. C., Mitchell, C. D., Applegate, K. R., Crony,
     E., Cabana, V. and Gjone, E. Scand. J. Clin. Lab. Invest. 35
     Suppl. 142: 31, 1975.
21. Glomset, J. A., Norum, K. R. and King, W. J. Clin. Invest.
     49: 1827, 1970.
22. Torsvik, H. Clin. Genet. 3: 188, 1972.
23. Utermann, G., Menzel, H. J. and Langer, K. H. FEBS Letters
     45: 29, 1974.
24. Anderson, D. W., Nichols, A. V. and Lindgren, F. T. Biochim.
     Biophys. Acta, in press.
25. Anderson, D. W., Nichols, A. V., Pan, S. S. and Lindgren,
     F. T. Atherosclerosis, accepted for publication.
26. Nichols, A. V., Gong, E., Blanche, P., Forte, T. M. and Anderson,
     D. W. Biochim. Biophys. Acta 446: 226, 1976.
27. Tall, A. R. and Small, D. M. Nature 265: 163, 1977.
28. Nichols, A. V., Forte, T. M., Gong, E. L., Blanche, P. J., and
     Nordhausen, R., in preparation.

ASSAY OF LIPOPROTEIN LIPASE IN POSTHEPARIN PLASMA
AFTER SUPPRESSION OF THE HEPATIC TRIGLYCERIDE LIPASE
WITH SODIUM DODECYL SULFATE

Marietta L. Baginsky, Flora Wu, and W. Virgil Brown

Department of Medicine
School of Medicine, University of California
La Jolla, California 92093

INTRODUCTION

Lipoprotein lipase (LPL), released by heparin into the circulating blood, is thought to originate from the capillary endothelium of a variety of tissues, including muscle and adipose
tissue (1). Much evidence suggests that this enzyme is of primary
importance in the degradation and removal of triglyceride from
plasma chylomicrons and very low density lipoproteins (2 - 4).
Assay of LPL in postheparin plasma has been complicated by the
presence of a second triglyceride lipase which is of hepatic origin
(5 - 7). The function of this hepatic triglyceride lipase (H-TGL)
in triglyceride clearance has not been elucidated, though it is
possible that it plays a role in the removal of partially degra-
dated lipoproteins following LPL action.

No substrate which is specific for LPL or H-TGL has been discovered, although marked differences in hydrolytic rates have been
noted with a variety of lipids, i.e. phospholipids (8,9), monoolein
(10,11) and palmitoyl-CoA (12). Therefore, determination of the
activity of these enzymes has been dependent on other characteristics such as differential inhibition by chemicals, antigenic
specificity, or chromatographic behavior. In the initial studies
of LPL from several different tissues, Korn observed that high
concentrations of NaCl (1M) or protamine sulfate markedly inhibited
this activity (13). Neither of these substances inhibited the
triglyceride lipase released when heparin was added to liver homo-
genates (5) or when liver was perfused with a heparin-containing
medium (14). This observation provided the basis for an assay in
which H-TGL could be determined in postheparin plasma after inhibition of LPL by protamine sulfate; LPL levels were calculated
as the difference between the values obtained before and after

treatment of postheparin plasma with protamine sulfate (15).
Direct assay of both LPL and H-TGL has been possible following
separation and partial purification of the two enzymes by heparin-
Sepharose affinity chromatography (16). In this method, both
enzymes are bound at low ionic strength (less than 0.4 M NaCl) to
a column of Sepharose beads containing covalently linked heparin.
The enzymes are then eluted with buffer containing different
concentrations of sodium chloride (0.75 M for H-TGL; 1.5 M for
LPL). This method permits the determination of each enzyme under
its optimal assay conditions. However, it is laborious and gives
falsely low values for LPL due to losses during chromatography.
Selective precipitation of H-TGL or LPL with antisera prepared to
the purified enzymes has provided a rapid and precise assay for
each of these enzymes in human postheparin plasma (17). This
method has the disadvantage of requiring generation of specific
antisera for H-TGL or LPL. Also, these antisera may be species
specific, thus limiting their use in comparative studies. Recent
experiments in our laboratory, using purified H-TGL and LPL, have
demonstrated marked differences in their stability and kinetic
properties when in the presence of certain detergents (12).
Sodium dodecyl sulfate (SDS) was found to inactivate H-TGL at
concentrations which were protective for LPL. This observation
has led to the development of a new method for the direct determination of LPL in postheparin plasma. This assay is simple and
rapid, and it can be easily established using commonly available
laboratory chemicals and equipment. H-TGL is also directly
determined from a separate aliquot of the postheparin plasma,
after inhibition of LPL by 0.75M NaCl.

METHODS AND MATERIALS

Postheparin plasma was obtained from normal and hyperlipoproteinemic volunteers after a 12-hour fast. Blood was collected
(into tubes without additional anticoagulant) 15 minutes after
the intravenous injection of heparin (Riker laboratories) at a
dose of 60 units/Kg body weight. Plasma aliquots were kept frozen
at -80°C until needed. Just prior to assay, the postheparin
plasma samples were thawed and, when required, precipitated fibrin
removed by a 30-min centrifugation at 270 x g. For H-TGL determination, the plasma sample was diluted with an equal volume of
0.2 M Tris-HCl buffer, pH 8.2, and assayed immediately afterwards.
Unless otherwise indicated, LPL was determined after pre-incubation (26°C) of the postheparin plasma diluted with an equal
volume of 100mM SDS in 0.2 M Tris-HCl, pH 8.2, for the time
periods indicated in the legend of figures. Pre-incubation conditions in an experiment using purified enzymes are given in the
legend of Fig. 1. Assays were done in triplicate in a final
volume of 0.5 ml. Hydrolysis of glyceryl trioleate by H-TGL was
measured at pH 8.8 in 0.2 M Tris-HCl - 0.75 M NaCl, no activator
serum added. LPL activity was measured at pH 8.2 in 0.2 M Tris-

HCl - 0.10 M NaCl buffer, plus 90 µl of serum as source of
activator apolipoprotein(s). The assay medium for both enzymes
contained per ml: 9 µmoles glyceryl tri (1-^{14}C)-oleate (SA = 0.088
µC/µmole) (emulsified by sonication in 5% gum arabic in 0.2 M Tris
of the appropriate pH), 100mg of fatty acid-free bovine serum
albumin (18) and 15mg of gum arabic. The assay was started by the
addition of 10 µl of the postheparin plasma solutions described
above. After 60 minutes of incubation at 28°C the reaction was
terminated and the released fatty acid determined as described by
Huttenen, et al. (19), with the following exception: elution of
the fatty acid from the resin was done by incubation for 90 minutes
at 70°C with 1ml of an equal mixture of 2 M KOH in methanol and
Triton X-100. Results were corrected for recovery as determined
using standard solutions of 1-^{14}C-oleic acid and were expressed
in µmoles free fatty acid liberated in an hour, per ml of postheparin plasma. Unlabeled glyceryl trioleate was from Nu-Check
Prep, Inc., glyceryl tri (1-^{14}C)-oleate from Dhom, SDS from East-
man (lot # 175X), Trizma Base and gum arabic were from Sigma Chemical Company.

RESULTS

The effect of SDS on purified H-TGL and LPL. - Previous
studies had indicated that purified H-TGL was inactivated (99%)
when pre-incubated with 1 mM SDS for 5 minutes at 26°C. LPL
activity, under the same conditions, was almost completely preserved
(12). The possible use of SDS for the selective inhibition of
H-TGL in postheparin plasma was investigated in a model system
prepared by the addition of pre-heparin plasma to each of the
purified enzymes (Fig. 1). In this experiment, complete inhibition
of H-TGL was also obtained in the presence of added plasma protein,
although higher concentrations of SDS were required than with
purified enzymes alone. The higher amount of SDS needed to obtain
the same effect is consistent with the known interaction of protein
and detergent on a weight to weight basis (20).
      As reported previously, LPL is inactivated when diluted in a
buffer without detergent (12). This was also found to be the case
here even in the presence of serum. However, LPL activity was fully
protected (and somewhat activated) when incubated in the presence
of SDS. LPL remained fully active at an SDS concentration of
25mM in the pre-incubation mixture (yielding 2mM in the assay
system). This concentration of SDS almost completely inactivated
H-TGL.
      Pre-incubation of postheparin plasma with SDS. - Using
postheparin plasma from normal donors, the lipase activity assayed
under conditions optimal for H-TGL declined progressively with
pre-incubation at increasing concentrations of SDS (Fig. 2).
Virtually no activity remained at 50mM. When these plasma samples
were assayed under conditions optimal for LPL (0.1M NaCl, plus
added serum) a decline in activity was also observed. However,

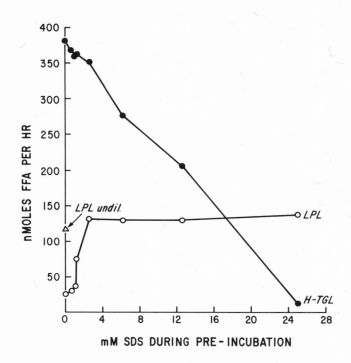

<u>Figure 1:</u> Effect of SDS on purified triglyceride lipases. H-TGL
and LPL were isolated from the postheparin plasma of a normal
volunteer by heparin-Sepharose chromatography (25). Both enzymes
in this experiment were used as obtained from the column (i.e.,
H-TGL in 0.005 M veronal buffer containing 0.75 M NaCl and LPL
in veronal buffer with 1.5 M NaCl, both of pH 7.0). Pre-incubation
mixtures for H-TGL contained: H-TGL solution (480 μg protein per
ml), veronal buffer - 1.5 M NaCl, normal serum and SDS (at double
the concentration indicated in the abscisa, in 0.2 M Tris-HCl of
pH 8.2) in the ratios (v/v) of 1:2:1:4. The pre-incubation
mixtures for LPL contained: LPL solution (132μg protein/ml),
veronal buffer -0.75 M NaCl, normal serum and SDS solutions (as
above) in the ratios of 2:1:1:4. Samples were pre-incubated for
10 minutes at 26°C, after which activity was determined as indicated in the Methods section, using 40 μl of enzyme solution.
Values given in the ordinate are expressed in nmoles FFA liberated
per hour per ml of undiluted enzyme solution.

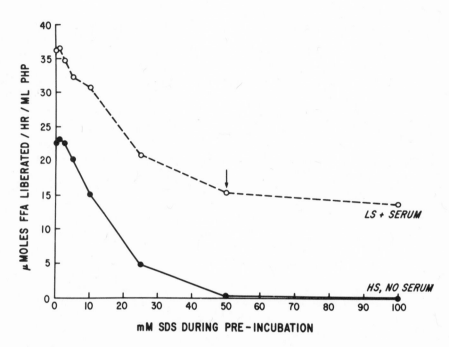

Figure 2: Effect of SDS concentration on H-TGL and LPL activities
of postheparin plasma. Postheparin plasma (500 µl) from a normal
human donor was mixed with an equal volume of SDS solution in 0.2
M Tris, pH 8.2,at concentrations to provide those indicated on the
abscisa. After 10 minutes at 26°C, assays for glyceryl trioleate
hydrolase activities were done as described in the Methods section.
The values indicated by LS + serum (0---0) were obtained using the
assay medium optimal for LPL; the curve indicated by HS, no serum
(●——●) was obtained with assays done under the conditions optimal
for H-TGL. The arrow indicates the SDS concentration (50mM)
chosen for all subsequent studies reported in this paper.

a plateau was reached which, in the example shown in Fig. 2, was
at approximately 1/2 the original value. This loss in activity is
consistent with inactivation of H-TGL, an enzyme that is also
determined (although sub-optimally) under the low salt condition.
The activity measured at concentrations of SDS of 50mM or above,
is thus believed to be due entirely to LPL.

Effect of duration of pre-incubation with SDS on enzyme
activities. Postheparin plasma samples from a normal subject and
from one patient each with Type I and Type V hyperlipoproteinemia
were compared during pre-incubation with 50mM SDS (Fig. 3). The
H-TGL activity (●——●) fell rapidly with a decline evident
even after 10 seconds of pre-incubation and, in the normal, no
residual activity was detectable after approximately 20 minutes
(Fig. 3A). When H-TGL activity had been suppressed to negligible
levels, activity assayed under low salt conditions (0 --- 0)
reached a constant value.

Postheparin plasma samples of Type I subjects assayed by
other methods have been found to be markedly deficient in LPL.
H-TGL is present, however, in subjects with this disorder (21-23).
This was confirmed in the postheparin plasma sample shown in Fig.
3B, using both heparin-Sepharose chromatography and immunochemical
techniques (data not given). The complete loss of glyceryl trioleate hydrolase activity by pre-incubation with SDS of this
plasma, known to be lacking in LPL, is again consistent with inactivation of H-TGL by SDS.

The experiment depicted in Fig. 3C demonstrates the effect
of pre-incubation with SDS of postheparin plasma from a subject
with Type V hyperlipoproteinemia. Activity measureable under the
assay conditions for H-TGL was still evident at 30 minutes in the
presence of 50mM SDS. This may be due, in part, to the very high
initial H-TGL activity found in this patient's plasma. This
slower rate of inactivation was also found in other hyperlipoproteinemic plasma samples with lower H-TGL levels. Thus, protection of H-TGL by the lipoproteins or lipid-detergent interactions could also be a cause for this slower inactivation. Extend-
ing the pre-incubation period to assure complete loss of H-TGL
activity raised, however, the possibility of producing serious
deterioration in LPL activity. Shown in Fig. 4 are results from
studies using postheparin plasma samples from subjects with Type
IV hyperlipoproteinemia and a normal individual, in which the pre-
incubation period was extended to 45 minutes. Whereas the
H-TGL activity approached negligible values, the activity measured
in low ionic strength and with serum activator was either stable
(Fig. 4A) or increased (Figs. 4B and 4C) with the more prolonged
pre-incubation in SDS.

The studies using postheparin plasma from two subjects with
Type V hyperlipoproteinemia is shown in Fig. 5. In these experiments, the pre-incubation with 50mM SDS was extended to one hour.
Note that the LPL activity did appear to reach a plateau at the
later time points. However, the prolonged survival of activity

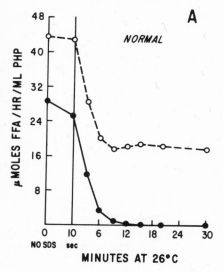

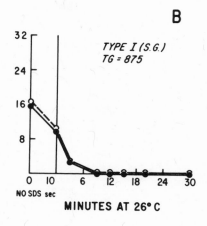

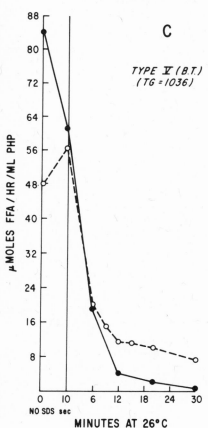

Figure 3: Glyceryl trioleate hydrolase activities remaining after pre-incubation with SDS. Postheparin plasma was pre-incubated with SDS (50mM final concentration), as described in the text. At the times indicated on the abscisa, aliquots were assayed under the conditions optimal for LPL (0---0) and for H-TGL (●—●). The plasma was obtained from a normal volunteer (Figure 3A) and subjects with Type I (Figure 3B) and Type V hyperlipoproteinemia (Figure 3C).

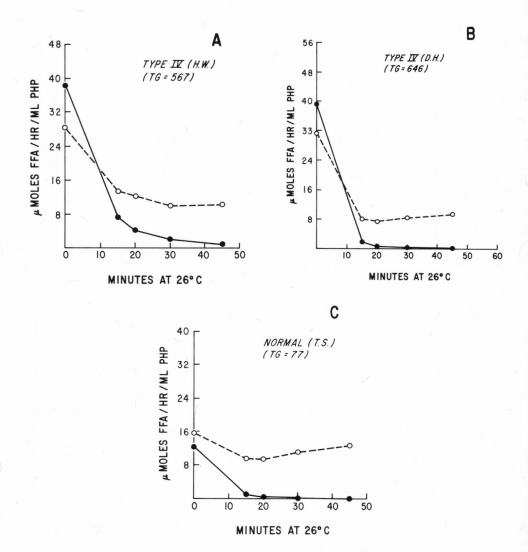

Figure 4: Glyceryl trioleate hydrolase activities remaining after 45 minutes of pre-incubation. Pre-incubation of postheparin plasma was carried out as described in the legend to Figure 3. At the times indicated on the abscisa assays were conducted under conditions optimal for LPL (0---0) and for H-TGL (●──●), as described in the Methods section. The values obtained with postheparin plasma from two subjects with Type IV hyperlipoproteinemia are shown in Figures 4A and 4B, and those from a normal subject in Figure 4C.

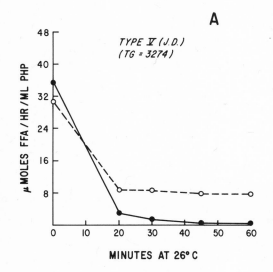

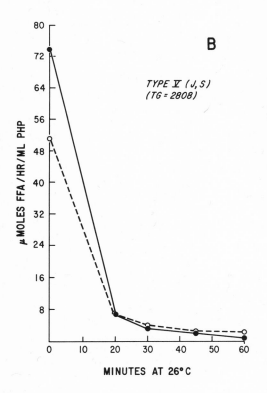

Figure 5: Effect of pre-incubation with SDS on glyceryl trioleate hydrolase activities of postheparin plasma from subjects with Type V hyperlipoproteinemia. The conditions of pre-incubation and assay were conducted as described in the legend of Figure 3. Subject A.D. has primary familial Type V hyperlipoproteinemia and subject J.S. has Type V hyperlipoproteinemia and insulin dependent diabetes mellitus. 0---0, assays done under conditions optimal for LPL; ●——●, assays done under conditions optimal for H-TGL.

in high salt concentrations continued to be a problem in certain patients as seen in Fig. 5B. Recently, pre-incubation experiments have been conducted for longer periods. The activity measured using assay conditions for LPL was maintained for at least 2 hours in these samples, while the H-TGL was completely inactivated. In the few postheparin plasma samples from Type V subjects examined thus far, it is clear that some have LPL activity within normal limits (Fig. 5A) while others have low activity (Fig. 5B). Of considerable interest is the very high H-TGL activity estimated in several of these plasma samples from subjects with Type V hyperlipoproteinemia.

DISCUSSION

Differing effects of low concentrations of detergents on the tertiary structure of a variety of proteins has been described (24). It is surprising, however, that two enzymes which appear to differ only in their carbohydrate side chains (25) could be so clearly distinguished in their response to exposure to SDS solutions. The stabilization of LPL may result, in part, from prevention of aggregation. The appearance of turbidity in solutions of purified LPL on reduction of the ionic strength by dilution has been repeatedly observed in our laboratory, and is consistent with a very high tendency for aggregation. When SDS is contained in the diluting buffer solutions, this turbidity is prevented and activity of the enzyme is fully maintained.

The selective inactivation of H-TGL is seen with both the purified enzyme and in plasma, over a broad range of lipoprotein values. With lipemic plasma samples, higher concentrations of SDS or more prolonged exposure is required to produce the same percentage of inactivation than with postheparin plasma from normal individuals.

The data presented in this article represents a preliminary report of the basis for a promising method for the direct assay of LPL in postheparin plasma. Conditions have been established empirically to completely inhibit all triglyceride hydrolase activity with the known assay characteristics of H-TGL. Using assay conditions defined for LPL, glyceryl trioleate hydrolase activity declined with the H-TGL inactivation, reaching a plateau which was stable for at least 2 hours of pre-incubation, under the conditions used. The remaining activity is believed to be due to LPL for the following reasons: 1) The purified LPL showed similar stability in the presence of pre-heparin plasma (Fig. 1) and no LPL activity was observed when this assay was done at high ionic strength (data not shown); 2) Similarly, the SDS-resistant activity of postheparin plasma was inhibited by 0.75 M NaCl (Fig. 2); and 3) Postheparin plasma from a subject with Type I hyperlipoproteinemia, (known to lack LPL) showed no glyceryl trioleate hydrolase activity after pre-incubation with SDS (Fig. 3B).

Initial results from a study currently underway suggest a

very close correlation between LPL, as assayed following immuno-
precipitation of H-TGL, and as assayed after complete inhibition of
the latter enzyme by pre-incubation with SDS. The values in the
latter method are 10-15% higher, perhaps reflecting some stabilization or activation of LPL in the presence of detergent.
Similar comparative studies with the LPL recovered from the
heparin-Sepharose affinity columns also showed a good correlation.
However, the latter values were 40-50% lower than those obtained
with the SDS method. The difference may reflect the summation of
LPL activation by SDS and the fractional loss of LPL during chromatography. On the other hand, values for H-TGL, were almost identical with the three methods.

   Using the same quantity of SDS, different degrees of inhibition of H-TGL were obtained with different batches of this detergent. Therefore, the concentration of SDS needed for complete
inhibition of H-TGL in postheparin plasma, as well as time of
pre-incubation required, have not yet been definitely set. This
is particularly true with plasma samples containing high lipoprotein levels.

   The acceptance of the values for LPL and H-TGL by the method
proposed in this communication is dependent on a clear demonstration of the precision of the procedure and accuracy of the
values obtained among the various forms of hyperlipoproteinemia.
The definition of accuracy must be made in a comparative sense
using the best method of assay now available. Such studies are
currently underway in our laboratory using the selective immunochemical precipitation of the two enzymes as the standard procedure for comparison.

ACKNOWLEDGEMENTS

   The excellent assistance of Mrs. Florence Casanada is
greatly appreciated.

This work was supported by the Specialized Center of Research on
Atherosclerosis (SCOR) as funded by the National Heart, Lung and
Blood Institute through grant #HL 14197-06.

REFERENCES

1. Robinson, D.S. and J.E. French. 1960. Heparin, the clearing factor lipase, and fat transport. Pharmacol. Rev. 12: 241-263.

2. Scow, R.O., J. Blanchette-Mackie and L.C. Smith. 1976. Role of capillary endothelium in the clearance of chylomicrons. A model for lipid transport from blood by lateral diffusion in cell membranes. Circulation Res. 39: 149-162.

3. Kopiang, I.P., A. Bensadoun, and M.W. Yang. 1976. Effect of an anti-lipoprotein lipase serum on plasma triglyceride removal. J. Lipid Res. 17: 498-505.

4. Jackson, R.L., J.D. Morrisett, and A.M. Gotto, Jr. 1976. Lipoprotein structure and metabolism. Physiological Rev. 56: 259-299.

5. La Rosa, J.C., R.I. Levy, H.G. Windmueller, and D.S. Fredrickson. 1972. Comparison of the triglyceride lipase of liver, adipose tissue, and post-heparin plasma. J. Lipid Res. 13: 356-363.

6. Greten, H., A.D. Sniderman, J.G. Chandler, D. Steinberg, and W.V. Brown. 1974. Evidence for the hepatic origin of canine postheparin plasma triglyceride lipase. FEBS Letters 42: 157-160.

7. Ehnholm, C., W. Shaw, H. Greten, W. Lengfelder and W.V. Brown. 1974. Separation and characterization of two triglyceride lipase activities from human postheparin plasma. In: Atherosclerosis III. Proceedings 3rd. International Symposium. G. Schetler and A. Winzer, Eds. Springer-Verlag, Berlin -Heidelberg -New York, NY, 557-560.

8. Ehnholm, C., W. Shaw, H. Greten and W.V. Brown. 1975. Purification from human plasma of a heparin-released lipase with activity against triglyceride and phospholipids. J. Biol. Chem. 250: 6756-6761.

9. Brown, W.V., W. Shaw, M. Baginsky, J. Boberg and J. Augustin. 1976. In: Lipoprotein Metabolism, H. Greten, Ed. Springer-Verlag, Berlin - Heidelberg - New York, NY, 2-6.

10. Greten, H., B. Walter and W.V. Brown. 1972. Purification of a human post-heparin plasma triglyceride lipase. FEBS Letters 27: 306-310.

11.   Nilsson-Ehle, P., T. Egelrud, P. Belfrage, T. Olivecrona and
        B. Borgström. 1973. Positional specificity of purified milk
        lipoprotein lipase. J. Biol. Chem. 248:6734-6737.

12.   Baginsky, M.L. and W.V. Brown. 1977. Differential characte-
        ristics of purified hepatic triglyceride lipase and lipo-
        protein lipase from human postheparin plasma. J. Lipid
        Res. 18: XX.

13.   Korn, E.D. 1955. Clearing factor, a heparin-activated lipo-
        protein lipase. I. Isolation and characterization of the
        enzyme from normal rat heart. J. Biol. Chem. 215: 1-14.

14.   Hamilton, R.L. 1965. Post-heparin plasma lipase from the
        hepatic circulation. Diss. Abstr. 26: 24.

15.   Krauss, R.M., H.G. Windmueller, R.I. Levy and D.S. Fredrick-
        son. 1973. Selective measurement of two different trigly-
        ceride lipase activities in rat postheparin plasma. J.
        Lipid Res. 14: 286-295.

16.   Boberg, J., J. Augustin, M.L. Baginsky, P. Tejada, and W.V.
        Brown. 1974. Quantitative determination of hepatic and lipo-
        protein lipase activities from human post-heparin plasma.
        Circulation 50: Suppl. III, 21.

17.   Huttunen, J.K., C. Ehnholm, P.K. Kinnunen and E.A. Nikkilä.
        1975. An immunochemical method for the selective measuremen†
        of two triglyceride lipases in human postheparin plasma.
        Clin. Chim. Acta 63: 335-347.

18.   Chen, R.F. 1967. Removal of fatty acids from serum albumin
        by charcoal treatment. J. Biol. Chem 242: 173-181.

19.   Huttenen, H.K., J. Ellingboe, R.C. Pittman, and D. Steinberg
        1970. Partial purification and characterization of hormone-
        sensitive lipase from rat adipose tissue. Biochim. Biophys.
        Acta 218: 333-346.

20.   Reynolds, J.A. and C. Tanford. 1970. The gross conformatio⁁
        of protein-sodium dodecyl sulfate complexes. J. Biol. Chem.
        245: 5161-5165.

21.   Krauss, R.M., R.I. Levy, and D.S. Fredrickson. 1974. Select
        ive measurement of two lipase activities in postheparin
        plasma from normal subjects and patients with hyperlipopro-
        teinemia. J. Clin. Invest. 54: 1107-1124.

22.   Greten, H., R. DeGrella, G. Klose, W. Rascher, J.L. deGennes,
       and E. Gjone. 1976. Measurement of two plasma triglyceride
       lipases by an immunochemical method: Studies in patients
       with hypertriglyceridemia. J. Lipid Res. 17: 203-210.

23.   Brown, W.V., M.L. Baginsky, and C. Ehnholm. (In press).
       Primary Type I and Type V hyperlipoproteinemia. In: Hyper-
       lipidemia: Diagnosis and Therapy. Rifkind, B. and R.I.
       Levy; (Eds.) Grune and Stratton, New York, NY.

24.   Steinhart, J. and J.A. Reynolds. 1969. Multiple equilibria
       in proteins. Academic Press, New York, NY; 234-302.

25.   Augustin, J., H. Freeze, J. Boberg, and W.V. Brown. 1976.
       Human postheparin plasma lipolytic activities. In: Lipo-
       protein Metabolism. H. Greten (Ed). Springer-Verlag, Berlin -
       Heidelberg - New York, NY: 7-12.

PHOSPHOLIPASE A$_2$ INTERACTION WITH VARIOUS PHOSPHOLIPIDS.

A MODEL FOR THE LIPOPROTEIN MOLECULE?

G.H. de Haas, A.J. Slotboom, H.M. Verheij, J.C. Vidal,
E.H.J.M. Jansen and H. Baartmans
Laboratory of Biochemistry, State University of Utrecht·
Transitorium 3, University Centre "De Uithof"
Padualaan 8, Utrecht, The Netherlands

INTRODUCTION

This NATO-workshop is devoted to the structure and function of
the lipoprotein molecule. Important building stones of these complexes are the so-called "apoproteins" and phospholipids. Considerable
progress has been made in the isolation and elucidation of the structure of proteins and phospholipids. One of the major problems is how
the lipids and proteins interact with each other on a molecular
level. Usually reconstitution experiments with pure apoproteins and
different types of phospholipids are monitored by calorimetric and
spectroscopic techniques such as fluorescence,ultraviolet difference
spectroscopy or NMR. However, no sensitive and easy bioassay is
available to control the structure of the reconstituted lipoprotein.
In addition, the tendency of several apoproteins to aggregate in
solution, complicates the interaction study. Therefore comparison
might be useful with a modelsystem in which a small water-soluble
protein molecule interacts with lipid aggregates, especially if the
interaction can be followed by a change in enzyme activity. In this
respect lipolytic enzymes are attractive: they are usually small,
rather stable proteins, which can easily be obtained in a pure state,
and, as we will see, the specific interaction of these proteins with
certain lipid-water aggregates is often accompanied by an exponen-
tial increase in enzyme activity.
    The purpose of the present study is to investigate the binding
forces between a lipolytic enzyme and its water-insoluble substrate
and to try to answer such questions as:
-which part of the protein molecule is involved in the interaction?
-should the interaction be defined as Adsorption, Anchoring or Penetration?

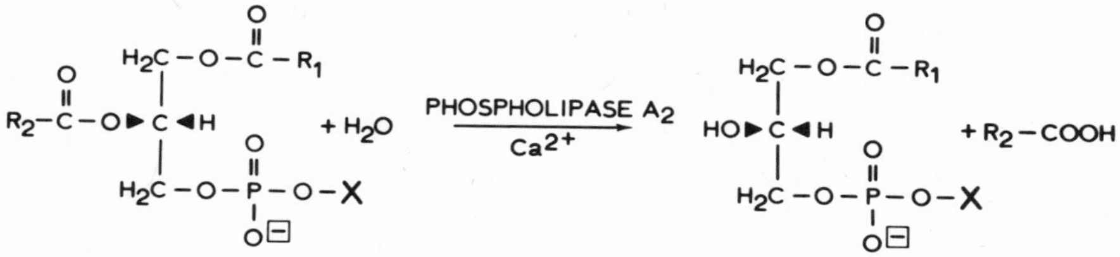

Fig. 1. *Reaction catalyzed by phospholipase A₂. X represents any of the naturally occurring moieties found in 3-sn-phospho-glycerides (e.g., choline, ethanolamine, glycerol, etc.). R₁ and R₂ are long alkyl chains.*

Fig. 2. *Revised primary structure of porcine pancreatic (pro)phos-pholipase A₂. (From Puijk et al. 10.)*

-is the interaction reversible?
-which parameters of the lipid-water interface control the interaction?

     Most of our work has been done with mammalian pancreatic phospholipases A_2 (EC 3.1.1.4). This enzyme has been isolated from ox, pig, horse and sheep pancreas and catalyzes the hydrolysis of the 2-acyl ester linkage in naturally occurring phosphoglycerides (1). (See Figure 1). The enzyme acts highly stereospecific and only 3-*sn*-phosphoglycerides are degraded whereas the stereoisomeric 1-*sn*-phosphoglycerides (D-α-phosphatides) are competitive inhibitors characterized by the same binding constant to the enzyme as the 3-*sn*-analogs (2). The pancreatic enzymes have an absolute requirement for Ca^{2+} ions which bind in a 1:1 molar ratio to the enzyme at the catalytic or high-affinity site. Ba^{2+} and Sr^{2+} ions bind to this site with a similar affinity constant as Ca^{2+}, but no hydrolysis takes place. Mg^{2+} ions do not bind to the enzyme (3). Sofar only a few transition metals such as Gd^{3+} and Tb^{3+} were found to replace Ca^{2+} with retention of some enzyme activity (4).

     The mammalian pancreatic phospholipases A_2 are all secreted by the pancreas in a zymogen form, probably to avoid uncontrolled degradation of membraneous material at the site of biosynthesis (5-9). The amino acid sequences of the porcine (10), bovine (11) and equine (12) phospholipase A_2 have been determined. Figure 2 shows the primary structure of the porcine zymogen, a protein whose X-ray structure has been determined recently by Drenth and colleagues (13). (Compare figures 10ᵃ and 10ᵇ).

     The zymogen consists of a single polypeptide chain of 131 amino acids cross-linked by 7 disulfide bridges. During conversion into the active enzyme by limited proteolysis with trypsin the N-terminal Arg-Ala¹ bond is split and the activation heptapeptide PyroGlu-Glu-Gly-Ile-Ser-Ser-Arg is liberated. After activation there is a conformational change which can be monitored by fluorescence spectroscopy of the unique Trp³ residue located at the N-terminal site (14). The newly generated α-NH_3^+ group of the N-terminal L-Ala¹, which has a pK of 8.3, most probably forms an internal ion-pair with a buried carboxylate function in the interior of the protein (15).

     The natural substrates of phospholipase A_2 are long-chain phospholipids which are insoluble in water and aggregate forming double-layered liposomes or vesicles. Taking into account that usually the lipolytic enzymes are small molecules and water-soluble, it is clear that lipolysis is a form of heterogeneous catalysis (17). The hatched area in figure 3 represents the lipid-water interface of the insoluble lipids. (= surface of the micelle, liposome, vesicle or monolayer at the air-water interface). The bulk aqueous phase contains the highly soluble enzyme E. Therefore lipolysis consists at least of two steps:

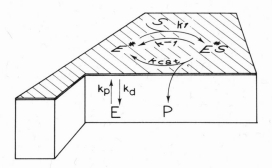

Fig. 3. *Model for the action of a soluble enzyme at an interface.*
(From Verger et al. 16.)

-the enzyme has to interact first with the lipid-water interface:
$E \rightleftarrows E^*$. We assume for the moment being that this interaction is
reversible and characterized by the rate constants k_p and k_d. Dependent on the extent of interaction this first step can be called
Adsorption, Anchoring or Penetration. Most probably the enzyme will
change its conformation in the new environment, denoted as E^*.
-Once the enzyme is located in the interface, we assume a "normal",
two-dimensional Michaelis-Menten equilibrium: one molecule of protein, being "dissolved in a sea of rapidly moving monomer lipid
molecules" (18-20), will react in a reversible way with a lipid
molecule S forming the activated complex E^*S. The E^*S complex de-
composes into product P and enzyme E^*. This model implies that the
activity of a lipolytic enzyme acting on a lipid aggregate is re-
gulated not only by the architecture of its catalytic site, but
will be strongly dependent also on the first step $E \rightleftarrows E^*$. Therefore it may be expected that a study of the kinetics of the enzyme
will provide valuable information on the interaction process $E \rightleftarrows E^*$.

As will be seen later on, it is difficult to use in kinetic
studies natural long-chain phospholipids as substrate. However, upon
acyl-chain shortening below nine carbon atoms, lecithins form micelles in stead of multilayered liposomes and enzymic degradation
of these micellar phospholipids can easily be quantitated by a number of techniques. As shown in Figure 4, using as substrate 1,2-di-
heptanoyl-*sn*-glycero-3-phosphorylcholine, both the active phospholipase A_2 (curve a) and the zymogen (curve b) follow normal Michaelis-
Menten kinetics up to the critical micellar concentration (CMC) of
the lecithin. This indicates that both proteins are able to hydrolyze monomeric substrate at comparable, but slow rates. Therefore,
it has to be concluded that in both proteins a similar active site
is present! (21). Kinetic difference becomes apparent upon substrate
aggregation: when micelles are formed, a tremendous increase in en-

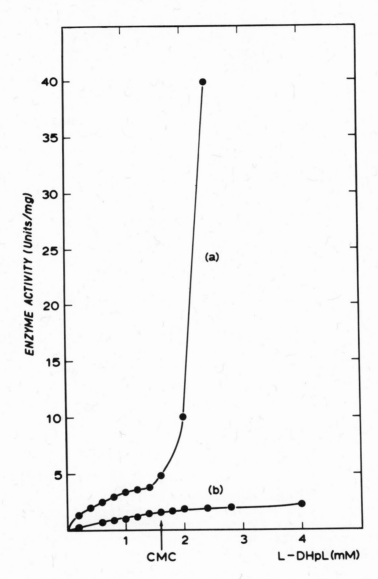

Fig. 4. *"Michaelis" curves showing the activity of phospholipase A$_2$ (curve a) and its zymogen (curve b) as a function of 1,2-diheptanoyl-sn-glycero-3-phosphorylcholine (= L-DHpL) concentration. Assay conditions: 0.5 mM NaAc, 0.1 M NaCl, pH 6.0, 40°. CMC stands for critical micellar concentration. (From Pieterson et al. 21.)*

zymic activity is observed for the active enzyme but not for the zymogen. Direct binding studies such as equilibrium gel filtration (21), ultraviolet difference spectroscopy (22), fluorescence measurements (22) and microcalorimetry (23) have shown that, in contrast to the active enzyme, the zymogen is not able to bind to lipid-water interfaces. The difference between both proteins has been attributed to the presence of a particular surface domain in the active enzyme through which it recognizes and binds to certain lipid-water interfaces (Fig. 3 E ⇄ E*). This site, the so-called Interface Recognition Site (IRS) is not present in the zymogen. From the above it can be concluded that there is a functional difference between the IRS and the active site area where *monomeric* substrates and Ca^{2+} are bound and hydrolysis occurs (*c.f.* Fig. 3).

Relative Location of the Active Site and IRS

As has been shown previously (24), porcine pancreatic phospholipase A_2 is irreversibly inactivated by *p*-bromophenacylbromide. This reagent specifically reacts with His^{48}. The fact that Ca^{2+} ions and monomeric substrate analogs give a nearly absolute protection against this modification seems to indicate that the reagent can be considered as an active site-directed inhibitor. Direct proof that the chemical modification is preceded by a reversible complex formation between enzyme and inhibitor could not be obtained for the porcine enzyme, probably because of a rather weak affinity of this protein for *p*-bromophenacylbromide. Inactivation experiments of the bovine enzyme as a function of the inhibitor* concentration, however, clearly show a saturating character (Figure 5), and a dissociation constant $K_i = 3$ mM was found at pH 5.5 (25). Also in this case it could be demonstrated that only His^{48} was modified and again protection was observed by Ca^{2+} ions and monomeric substrates. So it can be concluded that His^{48} is one of the active site residues of phospholipase A_2 and that binding of the essential ligands Ca^{2+} and monomeric lecithin takes place close to this residue. Although the *p*-bromophenacylbromide-inhibited phospholipase A_2 does not possess any enzyme activity, indicating a complete loss of active center, it has been demonstrated (2) that the modified protein still interacts with lipid-water interfaces. Mixtures of native and His^{48}-modified phospholipase A_2 have been studied in competition experiments for a micellar lipid-water interface by equilibrium gel filtration.

----

* Because of a higher solubility phenacylbromide was used in stead of *p*-bromophenacylbromide.

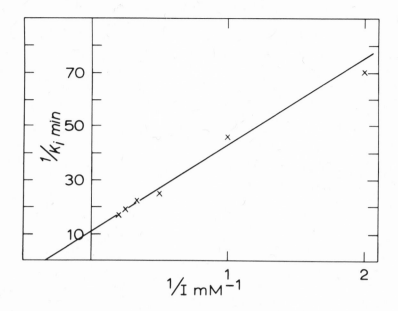

Fig. 5. *Double reciprocal plot of the inactivation velocity as a
function of the phenacylbromide concentration. Conditions:
0.1 M sodium cacodylate-HCl (pH 5.5) containing 0.1 M NaCl;
bovine phospholipase concentration 2 x 10⁻⁴ M. T = 25°C.
(From Dutihl, 25).*

Treatment of the data according to Best-Belpomme and Dessen (26)
yields the following equation:

$$\left[\frac{1}{\overline{LE}} - \frac{1}{\overline{LE_0}}\right] = \frac{K_M}{K_E} \cdot \frac{[M]}{[E]}$$

in which K_E and K_M are the affinity constants of the native and
modified enzyme, respectively. $[M]$ and $[E]$ are the concentrations
of unliganded modified and native phospholipase A₂, respectively.
LE is the fraction of sites combined with native phospholipase A₂,
LE_0 is the fraction of sites combined with native phospholipase A₂,
in the absence of modified phospholipase A₂.

Therefore a plot of $\left[\frac{1}{\overline{LE}} - \frac{1}{\overline{LE_0}}\right]$ as a function of $\frac{[M]}{[E]}$ should yield

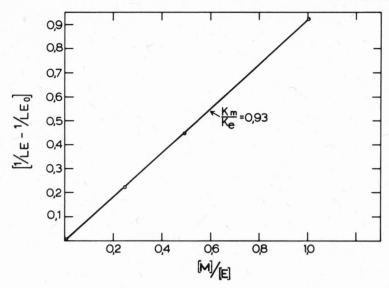

Fig. 6. *Plot of $[1/LE - 1/LE_0]$ vs $[M]/[E]$ according to the method of Best-Belpomme and Dessen (26) showing identical affinities of phospholipase A2 and p-bromophenacylbromide inhibited phospholipase A2 for micellair phospholipids (21).*

a straight line through the origin, the slope of which represents the ratio of the affinity constants. The result is given in **Figure** 6. It can be concluded that blocking of the active site in phospholipase A2 has no influence on the properties of the IRS. This implies that the IRS is not only functionally but also *topographically* distinct from the active site.

Which Amino Acids are involved in the IRS?

The intriguing question where the IRS is located on the enzyme surface has been investigated by a variety of techniques: Direct binding studies of phospholipase A2 (or specifically modified derivatives) with organized lipid-water interfaces using spectroscopic techniques indicated that Trp[3] and Tyr[69] are part of the IRS (22). Protection experiments against specific tryptic cleavage of the Arg[6]. Ser[7] linkage were performed in the presence of micellar substrate analogs and pointed to a possible role of Arg[6] in the IRS. Upon selective carboxymethylation of Met[8], the enzyme is no longer able to bind lipid-water interfaces (27). Equilibrium gel filtration and ultraviolet difference spectroscopy have been used to study the interaction process as a function of pH. Anchoring of the enzymes into micelles of substrate analogs appears to be governed by the protonation state of a single group with pK = 8.3. This residue has been identified as the α-NH$_3^+$ group of the N-terminal amino acid Ala^1 (21-22). In summary these results strongly suggest that the IRS embraces at least the rather hydrophobic N-terminal sequence Ala^1. Leu.Trp.Gln.Phe.Arg.Ser.Met^8 and Tyr^{69}. In the 3D-structure of the zymogen this latter residue is located close to the N-terminus. (*Cf.* Figures 10A, B).

The particular role of the N-terminal amino acid Ala^1 of phospholipase A$_2$ in the recognition process of lipid-water interfaces has been demonstrated using a number of phospholipase A$_2$ analogs in which the polypeptide chain has been elongated, shortened or in which native L-Ala^1 has been replaced by other amino acids (28-29). A number of these modified enzymes are compiled in Figure 7. Both

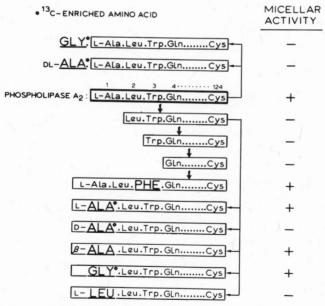

Fig. 7. *Scheme depicting selective substitution and addition of amino acids at the N-terminus of phospholipase A2.* ✱ *stands for 90%* ^{13}C *enriched amino acid (3-^{13}C D- and L-Ala; 1-^{13}C Gly). For experimental details see Slotboom and de Haas (28) and Slotboom* **et al.** *(29).*

chain elongation with an additional amino acid (Gly, DL-Ala)
and chain shortening yield proteins which have lost their
activity toward micellar substrates. A few substitutions in the
native polypeptide chain are not lethal to a functionally active
IRS: Gly^1 or β-Ala^1 in stead of L-Ala^1 and Phe^3 in stead of Trp^3.
However, the presence of the more bulky L-Leu^1 and even of the
stereoisomeric D-Ala^1 as N-terminal gives rise to a polypeptide
chain which no longer recognizes micellar substrates. It has to
be remarked that in these chemically modified phospholipase A_2
analogs the active center around His^{48} is still intact and
enzyme activity toward <u>monomeric</u> substrates is similar to that
of the native phospholipase A_2. These results strongly suggest
that the induction of the IRS requires a very exact juxtaposition
of the α-NH_3^+ group and its negatively charged countnerpartner.
Most probably the above modified phospholipases A_2 which have lost
their activity toward micellar substrates are not longer able to
bind to these lipid-water interfaces because the internal
saltbridge is impossible. An interesting consequence of the fact
that D-Ala^1-phospholipase A_2 has lost its affinity for micellar
substrates is the possibility to resolve the mixture of
diastereoisomers D-Ala^1-phospholipase A_2+L-Ala^1-phospholipase A_2
by "affinity chromatography"(29). As shown in figure 8^A,
DL-Ala^1-phospholipase A_2 elutes from Sephadex G_{100} at pH 6.0 as a
single symmetrical peak in a position corresponding to its molecular
weight of$\approx$14000. A simular elution experiment at pH 6.0 in which
the column was loaded with DL-Ala^1-phospholipase A_2 and micelles
of the substrate analog n-hexadecylphosphorylcholine is shown in
Figure 8B. The large lipid micelles elute at the breakthrough
volume of the column, but now two equal protein peaks are observed.
The L- Ala^1-phospholipase A_2 elutes with the micelles at the void
volume, whereas the D-Ala^1-phospholipase A_2, which has no affinity
for the lipid micelles, elutes at the same position as under 8A.
Because of the fact that the binding of phospholipase A_2 to lipid
micelles is governed by the pK of 8.3 of the α-NH_3^+ group, the pure
lipid- free L-Ala^1-phospholipase A_2 can be isolated from the void
volume fraction in Figure 8B by a repeated elution over the same
column now at pH 9.5 (Figure 8C). It has to be remarked that the
success of the above described separation technique of DL-Ala^1-
phospholipase A_2 into the pure D-&L-components is due to the peculiar
conformation of the N-terminal L-Ala^1 in native phospholipase A_2,
forming a salt-bridge with a buried carboxylate group.
Similar experiments using DL-Ala^{-1}-phospholipase A_2^* do not result in
any separation.

*DL-Ala^{-1}-phospholipase A_2 refers to phospholipase A_2 to which
DL-Ala has been covalently linked to the N-terminal amino acid
L-Ala^1 of the native enzyme.

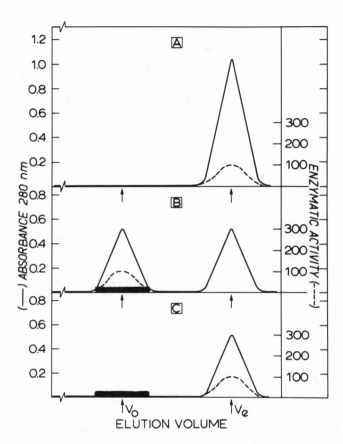

Fig. 8. *Schematic presentation of elution patterns of semisynthetic [DL-(3-^{13}C)-Ala$_1$]-phospholipase A$_2$ on a Sephadex G-100 column (60 × 1.2 cm). Experimental conditions: A, 4 mg of [DL-(3-^{13}C)-Ala$_1$]-phospholipase A$_2$, pH 6.0, 0.1 M acetate, 0.1 M NaCl; B, 4 mg of [DL-(3-^{13}C)-Ala$_1$]-phospholipase A$_2$ in the presence of 20 mg of n-hexadecylphosphorylcholine, pH 6.0, 0.1 M acetate, 0.1 M NaCl; C, fraction eluted at void volume (V$_o$) under B, pH 9.5, 0.1 M Tris-HCl, 0.1 M NaCl, 0.01 M EDTA. V$_o$ = void volume of the column; V$_e$ = elution volume of phospholipase A$_2$ (M. W. 14,000); ▬▬, position where micelles of n-hexadecylphosphorylcholine elute. Enzymatic activities (expressed in μequiv/ml) were determined as described by Nieuwenhuizen et al. (5) using a 1.5 fold higher sodium deoxycholate concentration. (From Slotboom et al. 29.)*

Role of Ca^{2+} in the interaction process

As discussed above the pancreatic phospholipases A_2 possess one "high-affinity" binding site for Ca^{2+}, located in the active center close to His[48]. This Ca^{2+} ion is absolutely required to enable the enzyme to hydrolyze monomeric substrates. Ca^{2+}, however, plays a second important role in the interaction between enzyme and organized lipid-water interfaces at alkaline pH. As shown in Figure 8B the enzyme binds to micellar interfaces at neutral or slightly acidic pH, but the complex dissociates at alkaline pH (Fig. 8C). It has been found by various direct binding techniques such as ultraviolet difference spectroscopy (22), fluorescence measurements (22), and microcalorimetry (23) that the interaction of phospholipase A_2 with lipid-water interfaces at basic pH is restored when specifically Ca^{2+} ions are present and that the complex remains intact up to high pH values. This is in agreement with the finding that the enzyme possesses a second optimum at pH 9.0 in the presence of relatively high Ca^{2+} concentrations (16,25) in addition to the optimum at pH 6.0 found at low Ca^{2+} concentration (30). These results have been tentatively explained by assuming that pancreatic phospholipase A_2 possesses at alkaline pH a second "low-affinity" Ca^{2+} binding site. Upon Ca^{2+} binding the protein undergoes a conformational change and the slightly altered polypeptide chain is able to anchor the lipid-water interface. Because micelle-binding requires an intact IRS, which is dependent on the internal salt bridge between the α-NH_3^+ group and a buried carboxylate, the effect of Ca^{2+} ions on the pK of the α-NH_3^+ group was studied. This was done by measuring the chemical shift of the resonance of phospholipase A_2 and some analogs containing a ^{13}C-enriched N-terminal amino acid as a function of pH. From the results given in Table 1, it can be seen that the pK of the α-NH_3^+ group of native phospholipase A_2 in the absence of Ca^{2+} ions is 8.4, which is in agreement with values determined previously by other methods. Upon addition of 50 mM Ca^{2+} the pK of the α-NH_3^+ group shifts to 9.0 and even to 9.3 when the Ca^{2+} concentration is increased to 100 mM. This increase in pK is rather specific for the N-terminal L-Ala[1] of native phospholipase A_2 and for Ca^{2+} ions. Mg^{2+} or Na^+ ions do not show any effect on this pK value, whereas the pK values of the α-NH_3^+ group of some enzymatically inactive phospholipase A_2 analogs having D-Ala as N-terminal amino acid or containing an extra amino acid (DL-Ala or Gly at the N-terminus show a drop from 8.3 to 7.6 upon addition of Ca^{2+}. These experiments strongly suggest that upon Ca^{2+} binding the salt bridge in native phospholipase A_2 moves to a much more hydrophobic surrounding. Due to this effect, which is even more pronounced in the presence of lipid-water interfaces, the salt bridge between the α-NH_3^+ group and the carboxylate, which triggers the IRS, remains intact up to higher pH values.
As regards the location of the "low-affinity" Ca^{2+} binding site in phospholipase A_2, no definite proposal can be given.

Table 1.   pK VALUES OF THE N-TERMINAL α-NH$_3^+$ GROUP OF PORCINE
PANCREATIC PHOSPHOLIPASE A$_2$ AND SOME ANALOGS

	Ca^{2+} (mM)			Mg^{2+} or NaCl (mM)
	0	50	100	100
1......124				
L-Ala*......Cys	8.4	9.0	9.3	8.4
D-Ala*......Cys				
DL-Ala*. L-Ala......Cys	8.3	7.6	-	7.8
Gly*. L-Ala......Cys				

*Stands for 90% ^{13}C enriched amino acid (3-^{13}C D- and L-Ala; 1-^{13}C
Gly). Data obtained from 90.5 mHz ^{13}C-NMR titration curves at 25°C
(1-2 mM protein solution in 50 mM Hepes buffer containing 10% D$_2$O).

In the absence of lipid-water interfaces the dissociation constant
has been estimated by various techniques to be about 20 mM at pH
7.5, which means that even high resolution X-ray of the enzyme most
probably will not be able to reveal its position. On the other hand,
the absence of this site in the zymogen (31) and the influence of
Ca^{2+} binding on the pK of the α-NH$_3^+$ group in phospholipase A$_2$ suggest that the second site will be located close to the N-terminal of
the polypeptide chain.

Is the enzyme-lipid interaction reversible?

The model shown in Figure 3 implies that the interaction of phospholipase A$_2$ with organized lipid-water interfaces is reversible. There
are a few reports, however, which might cast some doubt on this assumption:
Dervichian and coworkers (32) and Rietsch et al. (33) reported that
the desorption of lipolytic enzymes from a lipid monolayer at the
air-water interface is extremely slow. Dutilh (25) also favoured

a rather irreversible interaction of phospholipase A$_2$ with lipid-water interfaces and proposed an unfolding of the polypeptide chain at any interface. His argument is based mainly on elution profiles such as shown in Figure 8B, which would not be compatible with dissociation constants of 1 mM.

On the other hand, there are numerous experiments which can only be explained if the enzyme anchors the interface in a reversible way, for example:

- the determination of apparent dissociation constants between phospholipase A$_2$ and lipid micelles by ultraviolet difference spectroscopy (22), fluorescence (22) or microcalorimetry (23). The observed saturation characteristics are not compatible with an irreversible interaction.
- in kinetic studies dilution of the reaction mixture immediately results in lower enzymatic activities.
- rapid "desorption" of the enzyme from a micelle-enzyme complex upon a pH change from 8 to 9, as monitored by ultraviolet difference spectroscopy.
- the rapid endothermic heat effect observed in the microcalorimeter upon addition of EDTA to an enzyme-Ca^{2+}-micelle complex at pH 9.0. This heat effect equals the exothermic heat evolved in the formation of the E-Ca^{2+}-micelle complex in the absence of EDTA (23).

The reversibility of the anchoring process is observed not only with the somewhat artificial micellar substrate analogs, but also in the interaction between enzyme and liposomes of long-chain phospholipids. As was demonstrated by Op Den Kamp et al. (34), the pancreatic phospholipase A$_2$ can anchor to the closely packed bilayers of long-chain lecithins only at the transition temperature. Liposomes of dimyristoyl lecithin are hydrolyzed rapidly at 23°C, but below and above this temperature no hydrolysis is observed. Apparently the enzyme is not able to penetrate into the pure gel or liquid crystalline phase.

The reversibility of the penetration process is shown in Figure 9. 1,2-Di-tetradecyl-rac-glycero-3-phosphorylcholine, a non-hydrolyzable ether analog of dimyristoyl lecithin has a transition temperature as determined by Differential Scanning Calorimetry of 26.7°C. Ultraviolet difference spectroscopy of a mixture of vesicles of this phospholipid + phospholipase A$_2$ against lipid vesicles alone as a function of increasing temperature shows a sharp decrease of turbidity in the mixture at the transition temperature, indicating a change in bilayer structure caused by interaction with the enzyme (curve a). Above the transition temperature a homogeneous liquid-crystalline phase is formed, the enzyme is squeezed out and the "absorption" signal returns to the value found below the transition temperature. The curve is fully reversible and can be obtained also by cooling down the mixture from temperatures above the transition temperature. In the presence of the zymogen (curve b), which is unable to recognize the lipid-water interface, no change in absorption is obtained. It has to be remarked that at alkaline pH and in the absence of Ca^{2+} curve b is obtained also for the active phospholipase A$_2$.

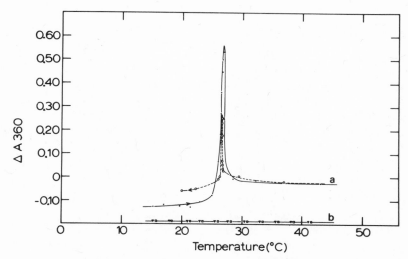

Fig. 9. *Change in turbidity at the transition temperature as mea-*
sured by ultraviolet difference spectroscopy at 360 nm upon
interaction of phospholipase A₂ with liposomes of 1,2-di-
tetradecyl-rac-glycero-3-phosphorylcholine at pH 8.5. ●,
phospholipase A₂ in the presence of 25 mM Ca²⁺ (starting
below the transition temperature); ○, phospholipase A₂ in
the presence of 25 mM Ca²⁺ (starting above the transition
temperature); △, prophospholipase A₂ in the presence of
25 mM Ca²⁺; □, phospholipase A₂ in the absence of Ca²⁺.

This observation is in agreement with the results given in Table 1
and demonstrates once more that the enzyme can penetrate lipid-water
interfaces at alkaline pH only after saturation of the second Ca²⁺
binding site.
Therefore, in our opinion there is no doubt that in general the
interaction of a lipolytic enzyme with an organized lipid-water
interface is reversible. This does not imply that the rate constants k_p and k_d in Figure 3 are of the same order of magnitude.
Especially monomolecular surface films at the air-water interface may
possess such a high interfacial energy (low surface pressure π) that
proteins are nearly irreversibly adsorbed or even denatured (32,33).
Therefore, kinetic studies of lipolytic enzymes using the monolayer
technique should preferably be done at higher surface pressures where
the substrate film possesses a low interfacial energy.
As regards elution profiles as shown in Figure 8B, they cannot be
used as an indication for irreversible binding of the enzyme to the
micelles.
The experimentally obtained dissociation constants between enzyme
and lipid micelles (22) of about 1 mM are apparent values expressed

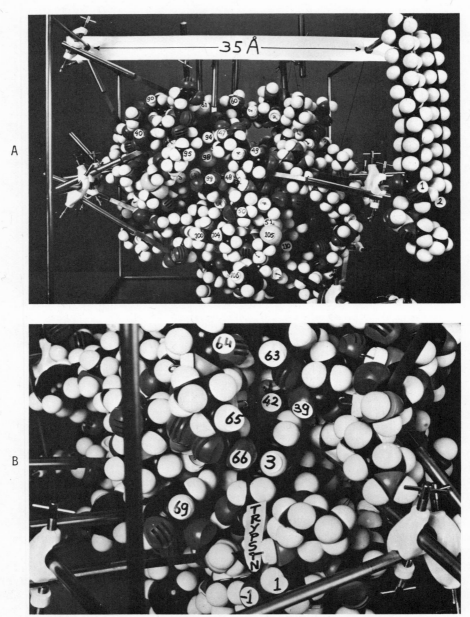

Fig. 10. *A survey of the active site region (fig. 10A) and of the N-terminus (fig. 10B) of the CPK molecular model of porcine pancreatic phospholipase A₂ (13).*

in the usual way as lipid monomers. It should be realized, however,
that the enzymes do not bind free monomers but interact with a lipid
aggregate. If one lipid micelle, consisting of about 200 monomers,
binds and transports one enzyme molecule, the dissociation constant
would not be 1 mM but 0.005 mM! In the latter case, elution profiles
as shown in Figure 8B can be expected.

Mechanism of action of pancreatic phospholipase A$_2$

In this final part we will not try to unravel the mechanism through
which the enzyme catalyzes the hydrolysis of monomeric substrates.
This subject has been discussed recently by Sigman and Mooser (35)
and Drenth et al. (13). What we would like to understand is how
membrane-like structures such as organized lipid-water interfaces
can control the activity of enzymes which by themselves are water-
soluble. What are the parameters of the interface which control the
activity of proteins anchored to the membrane?
The tremendous activation of phospholipase A$_2$ when acting upon certain organized lipid-water interfaces points to a very specific
interaction between the enzyme and the aggregated substrate (17).
Accelerating effects of 3 or 4 orders of magnitude compared with
turnover numbers found in the hydrolysis of monomeric substrates
cannot be explained by an increased substrate "concentration" only,
and we feel that the particular lipid-water interface, dependent on
its fine architecture, is able to induce, possibly in an "allosteric"
way, certain changes in the active site of the enzyme.
The IRS or supersubstrate binding site (36) through which the enzyme
anchors into the lipid phase has been shown to be not only functionally, but also topographically distinct from the active site. So
one may wonder how the substrate monomers present in the interface
arrive at the active center of the enzyme. The spatial separation
between the active center around His[48] and the IRS (N-terminal region of the polypeptide chain), suggested already by the two-dimensional structure given in Figure 2, can be seen also in Figures 10A
and 10B, which are photographs of opposite sites of a space filling
model of the zymogen of the porcine enzyme.
It might be argued, of course, that the X-ray structure given in
Figure 10 represents the zymogen of porcine pancreatic phospholipase
A$_2$ and not the active enzyme. However, even after splitting of the
activation peptide by trypsin, it is very difficult to imagine how
the newly formed N-terminal Ala[1] might form a salt bridge with a
carboxylate in the neighbourhood of His[48]. Bringing together of the
N-terminal part of the chain with the active center would require a
tremendous conformational change in the polypeptide chain. Taking
into account the highly compact structure of this protein, caused
by the presence of 7 disulfide bridges, such a conformational change
seems to be highly improbable.

Acknowledgments

The authors gratefully acknowledge the collaboration of many colleagues during the past years. In particular they want to express their thanks to Drs. R.Verger, C.E.Dutilh and H.Meyer for their important contributions to the present study. Thanks are due to Mr. W.C. Puijk and Mr. R. Dijkman for their excellent technical assistance.
We are indebted to Drs. P.S. de Araujo and M.Y. Rosseneu for providing results prior to publication and to Mr. G.G. van Deijl (Laboratory of Molecular Biology, University of UTRECHT) for making the photographs of the molecular model.
The authors thank Drs. P.W.M. van Dijck for the determination of the transitiontemperature of 1,2-ditetradecyl-rac-glycero-3-phosphorylcholine.
This study was partly carried out under the auspices of the Netherlands Foundation of Chemical Research (S.O.N.), and with financial aid from the Netherlands Organization for the Advancement of Pure Research (Z.W.O.).
The 90.5 MHz ^{13}C-NMR measurements were carried out at the SON NMR facility in Groningen (The Netherlands).

References

1) L.L.M. van Deenen and G.H. de Haas, Adv. Lipid Res., 2 (1964) 167

2) P.P.M. Bonsen, G.H. de Haas and L.L.M. van Deenen,
   Biochim. Biophys. Acta, 270 (1972) 364.

3) W.A. Pieterson, J.J. Volwerk and G.H. de Haas,
   Biochemistry, 13 (1974) 1439.

4) R.D. Hershberg, G.H. Reed, A.J. Slotboom and G.H. de Haas,
   Biochemistry, 15 (1976) 2268, 3902.

5) W. Nieuwenhuizen, H.B Kunze and G.H. de Haas,
   Methods Enzymol., 32^B (1974) 147.

6) C.E. Dutilh, P.J. van Doren, F.E.A.M. Verheul and G.H. de Haas,
   Eur. J. Biochem., 53 (1975) 91.

7) A. Evenberg, H. Meyer, H.M. Verheij and G.H. de Haas,
   Biochim. Biophys. Acta, 491 (1977) 265.

8) C. Figarella, F. Clemente and O. Guy,
   Biochim. Biophys. Acta, 227 (1971) 213.

9) B. Arnesjö, J. Barrowman and B. Borgström,
   Acta Chem. Scand., 21 (1967) 2897.

10) W.C. Puijk, H.M. Verhey and G.H. de Haas,
    Biochim. Biophys. Acta, 492 (1977) 254.

11) E.A.M. Fleer, H.M. Verhey and G.H. de Haas,
    Eur. J. Biochem., (1977) in press.

12) A. Evenberg, H. Meyer, W. Gaastra, H.M. Verhey and
    G.H. de Haas, J. Biol. Chem. 252 (1977) 1189.

13) J. Drenth, C.M. Enzing, K.H. Kalk and J.C. A. Vessies,
    Nature, 264 (1976) 373.

14) J.P. Abita, M. Lazdunski, P.P.M. Bonsen, W.A. Pieterson and
    G.H. de Haas, Eur. J. Biochem., 30 (1972) 37.

15) L.H.M. Janssen, S.H. de Bruin and G.H. de Haas,
    Eur. J. Biochem., 28 (1972) 156.

16) R. Verger, M.C.E. Mieras and G.H. de Haas,
    J. Biol. Chem., 248 (1973) 4023.

17) R. Verger and G.H. de Haas,
    Ann. Rev. Biophys. Bioeng. 5 (1976) 77.

18) D. Chapman, Ann. N.Y. Acad.Sci., 137 (1966) 745.

19) D. Chapman, P. Byrne and G.G. Shipley,
    Proc. Roy. Soc., 290^A (1966) 115.

20) S.J. Singer and G.L. Nicolson
    Science, 175 (1972) 720.

21) W.A. Pieterson, J.C. Vidal, J.J. Volwerk and G.H. de Haas,
    Biochemistry, 13 (1974) 1455.

22) M.C.E. van Dam-Mieras, A.J. Slotboom, W.A. Pieterson and
    G.H. de Haas, Biochemistry, 14 (1975) 5387.

23) P.S. de Araujo, M.Y. Rosseneu, A.J. Slotboom and G.H. de Haas,
    Manuscript in preparation.

24) J.J. Volwerk, W.A. Pieterson and G.H. de Haas,
    Biochemistry, 13 (1974) 1446.

25) C.E. Dutilh, Ph.D. Thesis, 1976, State University UTRECHT,
    The Netherlands.

26) M.Best-Belpomme and P. Dessen, Biochimie 55 (1973) 11.

27) F. van Wezel, A.J. Slotboom and G.H. de Haas,
    Biochim. Biophys. Acta, 452 (1976) 101.

28) A.J. Slotboom and G.H. de Haas, Biochemistry, 14 (1975) 5394.

29) A.J. Slotboom, M.C.E. van Dam-Mieras and G.H. de Haas,
    J. Biol. Chem., 252 (1977) 2948.

30) G.H. de Haas, P.P.M. Bonsen, W.A. Pieterson and
    L.L.M. van Deenen, Biochim. Biophys. Acta, 239 (1971) 252.

31) A.J. Slotboom, M.C.E. van Dam-Mieras, E.H.J.M. Jansen,F. Pattus,
    H.M. Verhey and G.H. de Haas,CNRS International Colloquium on
    Enzymes of Lipid Metabolism, April 1977, Strasbourg (France).
    Eds. P. Mandel, L. Freysz and S. Gatt, Plenum Press, New York,
    in press.

32) D.G. Dervichian, C. Préhu and J.P. Barque,
    C.R. Acad. Sci. Paris,276 (1973) 839.

33) J. Rietsch, F. Pattus, P. Desnuelle and R. Verger,
    J. Biol. Chem., 252 (1977) 4313.

34) J.A.F. Op Den Kamp, M. Th. Kauerz and L.L.M. van Deenen,
    Biochim. Biophys. Acta, 406 (1975) 169.

35) D.S. Sigman and G. Mooser, Ann. Rev. Biochem., 44 (1975) 913.

36) H. Brockerhoff and R.G. Jensen,
    Lipolytic Enzymes, (1974),p.17 New York: Academic Press.

Molecular Variation and Pathology

FAMILIAL LIPOPROTEIN DEFICIENCY

G. ASSMANN

KLINISCHE CHEMIE, UNIVERSITÄT KÖLN

5 KÖLN, JOSEPH-STELZMANN-STR.

WEST GERMANY

There are three genetically determined disorders in which one or more of the lipoprotein families are absent from plasma or their concentrations are extremely low: abetalipoproteinemia, hypobetalipoproteinemia, and Tangier disease (familial high density lipoprotein deficiency). Hypocholesterolemia is a common manifestation of all these diseases, which most easily can be distinguished on the basis of clinical symptoms and lipoprotein patterns.

Abetalipoproteinemia
Abetalipoproteinemia (Apo B deficiency) is a rare inborn error of lipid metabolism, characterized in the affected homozygote by the complete absence of chylomicrons, very low density lipoproteins (VLDL), and low density lipoproteins (LDL). The disease is associated with clinical evidence of fat malabsorption, retinitis pigmentosa, neuropathic ataxia, and acanthocytic red blood cells. The syndrom was first described in 195o in an 18-years-old-Jewish girl by Bassen and Kornzweig (Bassen-Kornzweig-syndrom). The characteristic lipo-protein changes in abetalipoproteinemia were discovered in 196o. As of today, approximately 35 cases have been reported. Comprehensive review articles on the clinical symptoms, the lipoprotein abnormality, the mode of inheritance, and the pathophysiology of this disease appeared in the literature.

Review articles

Schwartz,J.F.,Rowland,L.P.,Eder,H.,Marks,P.A.,Ossermann,
E.F.,Hirschberg,E.,Anderson,H. (1963)
Bassen-Kornzweig syndrome: deficiency of serum ß-lipoprotei
Arch.Neurol. 8,438

Isselbacher,K.J.,Scheig,R.,Plotkin,G.R.,Caufield,J.B.(1964)
Congenital ß-lipoprotein deficiency: an hereditary disorder
involving a defect in the absorption and transport of
lipids.
Medicine 43,347

Wolff,O.H. (1965)
A-beta-lipoproteinemia
Ergebn.Inn.Med.Kinderheilk.23,19o

Farguhar,J.W.,Ways,P. (1966)
Abetalipoproteinemia
The Metabolic Basis of Inherited Disease, 2nd ed.,
J.B.Stanbury,J.B.Wyngaarden,D.S.Fredrickson,eds.,
McGraw-Hill,New York

Kahlke,W. (1967)
A-ß-lipoproteinemia
Lipids and Lipidoses, ed. by G.Schettler
Springer Verlag, New York

Fredrickson,D.S., Gotto,A.M., Levy,R.I. (1972)
Familial lipoprotein deficiency
The Metabolic Basis of Inherited Disease, 3rd ed.,
J.B.Stanbury,J.B.Wyngaarden,D.S.Fredrickson,eds.,
McGraw-Hill,New York

Kayden,H.J. (1972)
Abetalipoproteinemia
Amer.Rev.Med.23, 285

Kostner,G.(1976)
ApoB-deficiency (Abetalipoproteinemia): A model for
studying the lipoprotein metabolism
Lipid Absorption: Biochemical and clinical aspects
K.Rommel,H.Goebell,R.Böhmer, eds.,MTP Press

Herbert,P.N., Fredrickson,D.S.(1976)
The Hypobetalipoproteinemias
Fettstoffwechsel,Handbuch der inneren Medizin.Fünfte
Auflage,Band 7,Stoffwechselkrankheiten, 4. Teil
G.Schettler,H.Greten,G.Schlierf,D.Seidel, eds.,
Springer-Verlag,New York

lipoprotein abnormalities in abetalipoproteinemia

Levy,R.I.,Fredrickson,D.S.,Laster,L. (1966)
The lipoproteins and lipid transport in abetalipoproteinemia
J.Clin.Invest. 45, 531

Jones,J.W., Ways,P. (1967)
Abnormalities of high density lipoproteins in abetalipo-
proteinemia.
J.Clin.Invest. 46, 1151

Lees,R.S. (1967)
Immunological evidence for the presence of B protein
(apoprotein of ß-lipoprotein) in normal and abetalipopro-
teinemic plasma
J.Lipid Res. 8, 396

Lees,R.S., Ahrens,E.H.,Jr. (1969)
Fat transport in abetalipoproteinemia
New Engl. J. Med. 28o, 1261

Gotto,A.M., Levy,R.I., John,K., Fredrickson,D.S. (1971)
On the nature of the protein defect in abetalipopro-
teinemia
New Engl. J. Med. 284, 813 (1971)

Forte,T., Nichols,A.V. (1972)
Application of electron microscopy to the study of
plasma lipoprotein structure
Advanc.Lipid Res. 1o, 1

Kostner,G., Holasek,A., Bohlmann,H.G., Thiede,H. (1974)
Investigation of serum lipoproteins and apoproteins in
abetalipoproteinemia
Clin.Sci.molec.Med. 46, 457

Scanu,A.M., Aggerbeck,L.P., Kruski,A.W., Lim,C.T.
Kayden,H.J. (1974)
A study of the abnormal lipoproteins in abetalipopro-
teinemia
J.Clin.Invest. 53, 44o

Hypobetalipoproteinemia

Familial hypobetalipoproteinemia is considered a genetic
disease distinct from abetalipoproteinemia. LDL are abnormally low but present; the disease appears to be
transmitted as an autosomal dominant trait. The diagnostic
criteria include the detection of a similar pattern in
a first-degree relative, and the absence of a disease to

which hypobetalipoproteinemia may be secondary. Moderate
hypocholesterolemia and hypotriglyceridemia without
acanthocytosis, fat malabsorption, and neuromuscular
disease occur in most patients studied so far. The
homozygous form of familial hypobetalipoproteinemia is
clinically and chemically indistinguishable from abetalipoproteinemia.

case reports

Salt,H.B., Wolff,O.H., Lloyd,J.K., Fosbrooke,A.S.,
Cameron,A.H., Hubble,D.V. (1960)
On having no beta lipoprotein. A syndrome comprising
a-beta-lipoproteinemia, acanthocytosis, and steatorrhea
Lancet II, 325

Van Buchem,F.S.P., Pol,G., de Gier,J., Böttcher,C.J.F.,
Pries,C. (1966)
Congenital ß-lipoprotein deficiency
Amer.J.Med. 4o, 794

Mars,H., Lewis,L.A., Robertson,A.L.,Jr., Butkus,A.,
Williams,G.H.,Jr. (1969)
Familial hypo-ß-lipoproteinemia. A genetic disorder
of lipid metabolism with nervous system involvement
Amer.J.Med. 46, 886

Richet,G., Durepaire,H., Hartmann,L., Ollier,M.P.,
Polonovski,J., Maitrot,B. (1969)
Hypolipoproteinemie familiale asymptomatique predominant
sur les beta-lipoproteines
Presse med. 77, 2o45

Levy,R.I., Langer,T., Gotto,A.M., Fredrickson,D.S. (197o)
Familial hypobetalipoproteinemia, a defect in lipoprotein
synthesis.
Clin.Res. 18, 539

Mawatari,S., Iwashita,H., Kuroiwa,Y. (1971)
Familial hypo-ß-lipoproteinemia
J.neuro.Sci. 16, 93

Fosbrooke,A., Choksey,S., Wharton,B. (1973)
Familial hypo-ß-lipoproteinemia
Arch.Dis.Childh. 48, 729

Brown,B.J., Lewis,L.A., Mercer,R.D. (1974)
Familial hypobetalipoproteinemia: report of a case with
psycho-motor retardation
Pediatrics 54, 11

Aggerbeck,L.P., McMahon,J.P., Scanu,A (1974)
Hypobetalipoproteinemia: Clinical and biochemical
description of a new kindred with Friedreich`s ataxia
Neurology 24, 1o51

Cottrill,C., Glueck,C.J., Leuba,V., Millett,F.,
Puppione,D., Brown,W.V. (1974)
Familial homozygous hypobetalipoproteinemia
Metabolism 23, 779

Biemer,J.J., McCammon,R.E. (1975)
The genetic relationship of abetalipoproteinemia and
hypobetalipoproteinemia: a report of the occurence of
both diseases within the same family
J.Lab.clin.Med. 85, 556

Naito,H.K., Lewis,L.A. (1975)
Serum lipoproteins and lipids of familial hyper-,hypo-,
or normo-ß-lipoproteinemic subjects during pregnancy
Clin.Chem. 21, 99o

Tamir,I., Levtow,O., Lotan,D., Lequin,C., Heldenberg,D.,
Werbin,B. (1976)
Further observations on familial hypobetalipoproteinemia
Clin.Genet. 9, 149

Glueck,C.J., Tsang,R.C., Mellies,M.J., Fallat,R.W.,
Steiner,P.M. (1976)
Neonatal familial hypobeta-lipoproteinemia
Metabolism 25, 611

Glueck,C.J., Gartside,P., Fallat,R.W., Sielski,J.,
Steiner,P.M. (1976)
Longevity syndromes: familial hypobeta- and familial
hyperalphalipoproteinemia
J.Lab.Clin.Med. 88, 941

Tangier disease

Familial hypoalphalipoproteinemia (Tangier disease)
was discovered in 196o by Fredrickson et al. and given
its name from Tangier island which lies off the coast
of Virginia in Chesapeake Bay and is the home of the
original patients. To date, approximately 25 patients
have been discovered. The clinical manifestations
include: low plasma cholesterol and normal or elevated
triglyceride concentrations; absence of the usual high
density lipoproteins and abnormal composition of other
lipoproteins in plasma; distinctive enlargement and
coloration of the tonsils; frequently peripheral neuro-

pathy; cholesteryl ester deposition in tissues accompanied
by enlargement of spleen,liver,or lymph nodes, alterations
in the cornea, intestinal mucosa, and other tissues.

Review articles

Fredrickson,D.S., Altrocchi,P.H. (1962)
Tangier disease (Familial cholesterolosis with high-
density lipoprotein deficiency
Cerebral Sphingolipidoses,A symposium on Tay Sachs
Disease and allied disorders, S.M.Aronson,B.W.Volk,eds.
Academic Press, New York

Fredrickson,D.S. (1966)
Familial high-density lipoprotein deficiency: Tangier
disease
The Metabolic Basis of Inherited Disease, IInd. ed.
J.B.Stanbury,J.B.Wyngaarden,D.S.Fredrickson,eds.
McGraw-Hill, New York

Fredrickson,D.S.,Gotto,A.M., Levy,R.I. (1972)
Familial lipoprotein deficiency
The Metabolic Basis of Inherited Disease, IIIrd. ed.
J.B.Stanbury,J.B.Wyngaarden,D.S.Fredrickson
McGraw-Hill, New York

Assmann,G. (1976)
Tangier Krankheit
Fettstoffwechsel. Handbuch der inneren Medizin. Fünfte
Auflage. Band 7. Stoffwechselkrankheiten. 4.Teil
G.Schettler,H.Greten,G.Schlierf,D.Seidel, eds.
Springer Verlag, New York

case reports

Fredrickson,D.S., Altrocchi,P.H., Avioli,L.V., Goodman,D.S.,
Goodman,H.C. (1961)
Tangier disease. Combined clinical staff conference at
the National Institutes of Health.
Ann.Intern.Med. 55, 1o16

Fredrickson,D.S., Young,O., Shiratori,T., Briggs,N.(1964)
The inheritance of high density lipoprotein deficiency
(Tangier disease)
J.Clin.Invest. 43, 228

Hoffman,H.N., Fredrickson,D.S. (1965)
Tangier disease (familial high density lipoprotein
deficiency). Clinical and genetic features in two adults.
Am.J.Med. 39, 582

Kocen,R.S., Lloyd,J.K., Lascelles,P.T., Fosbrook,A.S.
Williams,D. (1967)
Familial alpha-lipoprotein deficiency (Tangier disease)
with neurological abnormalities
Lancet I, 1341

Kummer,H., Laissue,J., Spiess,H., Pflugshaupt,R., Bucher,U
(1968)
Familiäre Analphalipoproteinämie (Tangier Krankheit)
Schweiz.med.Wschr. 98, 4o6

Huth,K., Kracht,J., Schoenborn,W., Fuhrmann,W. (197o)
Tangier-Krankheit (Hyp-alpha-lipoproteinämie)
Dtsch.med.Wschr. 95, 2357

Clifton-Bligh,P., Nestel,P.J., Whyte,H.M. (1972)
Tangier disease: report of a case and studies of lipid
metabolism
New Engl.J.Med. 286, 567

Greten,H., Hannemann,T., Gusek,W., Vivell,O. (1974)
Lipoproteins and lipolytic plasma enzymes in a case of
Tangier disease
New Engl.J.Med. 291, 548

Brook,J.G., Lees,R.S., Yules,J., Cusack,B. (1977)
Tangier disease in New England-report of a case
JAMA (J.Am.Med.Assoc.) in press

lipoprotein abnormalities

Lux,S.E., Levy,R.I., Gotto,A.M., Fredrickson,D.S. (1972)
Studies on the protein defect in Tangier disease. Isolation
and characterization of an abnormal high density lipoprotein
J.Clin.Invest. 51, 25o5

Kostner,G., Holasek,A., Schoenborn,W., Fuhrmann,W. (1972)
Immunchemische Untersuchung und isoelektrische
Fokussierung des Serums eines Patienten mit Tangier
Krankheit
Clin.chim.Acta 38, 155

Utermann,G., Menzel,H.J., Schoenborn,W. (1972)
Plasma lipoprotein abnormalities in a case of primary
high-density lipoprotein (HDL) deficiency
Clin.Genetics 8, 258

Assmann,G., Smootz,E., Adler,K., Capurso,A., Oette,K.
(1977)
The lipoprotein abnormality in Tangier disease. Quantitation of A apoproteins
J.Clin.Invest. 59, 565

Assmann,G., Herbert,P.N., Fredrickson,D.S., Forte,T.
(1977)
Isolation and characterization of an abnormal high density
lipoprotein in Tangier disease
J.Clin.Invest., in press

Assmann,G., Simantke,O., Schaefer,H.E., Smootz,E. (1977)
Characterization of high density lipoproteins in patients
heterozygous for Tangier disease
J.Clin.Invest., in press

tissue abnormalities

Waldorf,D.S., Levy,R.I., Fredrickson,D.S. (1967)
Cutaneous cholesterol ester deposition in Tangier disease
Arch.Derm. 95, 161

Engel,W.K., Dorman,J.D., Levy,R.I., Fredrickson,D.S.
(1967)
Neuropathy in Tangier disease. Alpha-lipoprotein deficiency
manifesting as familial recurrent neuropathy and intestinal lipid storage
Arch.Neurol. (Chic.) 17, 1

Shacklady,M.M., Djardjouras,E.M., Lloyd.J.K.(1968)
Red-cell lipids in familial alpha-lipoprotein deficiency
(Tangier disease)
Lancet II, 151

Spiess,H., Ludin,H.P., Kummer,H. (1969)
Polyneuropathie bei familiärer Analphalipoproteinämie
Nervenarzt 4o, 191

Kracht,J., Huth,K., Schoenborn,W., Fuhrmann,W. (197o)
Hypo-alpha-lipoproteinämie (Tangier Disease)
Verhdl.Dtsch.Ges.Path. 54, 355

Bale,P.M., Clifton-Bligh,P., Benjamin,B.N.P., Whyte,H.M.
(1971)
Pathology of Tangier disease
J.Clin.Pathol. (Lond.) 24, 6o9

Kocen,R.S., King,R.H.M., Thomas,P.K., Haas,L.F. (1973)
Nerve biopsy findings in two cases of Tangier disease
Acta Neuropath. 26, 317

Ferrans,V.J., Fredrickson,D.S. (1975)
The pathology of Tangier disease
Amer.J.Pathol. 78, 1o1

Gheorghiu,Th., Assmann,G., Schaefer,H.E. (1976)
Endoscopic findings in Tangier disease.
Endoscopy 8, 164

Katz,S.S., Small,D.M., Brook,J.G., Lees,R.S. (1977)
The storage lipids in Tangier disease. A physical chemical
study
J.Clin.Invest. 59, 1o45

FAMILIAL HYPERCHOLESTEROLAEMIA

G.R. THOMPSON

Medical Research Council Lipid Metabolism Unit

Hammersmith Hospital, London, England

Familial Hypercholesterolaemia (FH), or familial type II hyperlipoproteinaemia (1), is inherited in an autosomal dominant manner and is characterised by abnormally high levels of low density lipoprotein (LDL or β-lipoprotein) in plasma from birth. It has been estimated that FH affects 1:500 of the population of the United States of America and is responsible for 1:20 of all cases of myocardial infarction (2). Thus the disease is one of the commoner disorders inherited by humans and one which exerts a considerable toll in terms of premature morbidity and mortality. The current interest in attempting to reduce the prevalence of atherosclerosis in Western Countries has focussed attention on FH as a means of proving or disproving the lipid hypothesis of atherosclerosis. If the well-established relationship between hypercholesterolaemia and coronary heart disease is neither fortuitous nor due to some unknown third factor, then hypercholesterolaemia must either be a cause or a consequence of coronary heart disease. Since the hypercholesterolaemia of FH antedates the coronary heart disease by several years, it seems reasonable that the former causes the latter, and not vice versa. If so, effective reduction of hypercholesterolaemia in this monofactorial situation should result theoretically in regression of atherosclerosis, assuming that the atheromatous lesions in FH respond in an analagous manner to those which have been induced experimentally in monkeys and pigs (3, 4). Thus FH merits study not only as a disease entity in its own right but also because it provides a unique opportunity to determine whether lipid-lowering therapy is likely to be beneficial in the prevention and regression of premature coronary heart disease in the population at large (1). As Robert Frost once said: 'Nature is always hinting at us. It hints over and over again. And suddenly we take the hint'.

CLINICAL FEATURES

Homozygous inheritance of the FH gene, which is fortunately rare, is characterised by the appearance during childhood of cutaneous and tendinous xanthomata and the premature onset of aortic and coronary atheroma in association with markedly elevated levels of LDL in plasma.   Sudden death before the age of 30 is the rule rather than the exception (5), and is attributable to acute coronary insufficiency secondary to coronary ostial stenosis (6).   In life aortic involvement is manifested by an aortic systolic murmur without an ejection click, a gradient across the aortic valve and narrowing of the aortic root on aortography (7, 8).   At post-mortem both the aortic valve and especially the supravalvar region of the aorta are infiltrated, thickened and narrowed by atheroma (9, 10). Hisologically the lesions closely resemble 'ordinary' atheroma but contain more cholesterol and more numerous foam cells.   Atheroma is also found distally throughout the aorta, although to a lesser extent than proximally (11), and in the carotid, basilar and pulmonary arteries.

Heterozygotes usually present in early to mid-adult life with ischaemic symptoms such as angina, intermittent claudication or even vertebro-basilar insufficiency (12).   Their degree of hypercholesterolaemia is less severe than in homozygotes and they do not exhibit the cutaneous planar xanthomata which are characteristic of the latter.   The majority of heterozygotes have tendon xanthomata by the age of 30 (13), commonly affecting the Achilles tendons and the extensor tendons of the hands.   Corneal arcus is also common but xanthelasma less so (14).   FH is more often associated with a type IIa phenotype than with a type IIb, although the proportion of patients exhibiting the latter rises with age (13).   Up to 30 % of heterozygotes have an aortic systolic murmur (15) but autopsy studies reveal only supravalvar aortic atheroma, without involvement of the valve itself (11).   Coronary atheroma is common, with an ominous predilection for the main stem of the left coronary artery (16).

MOLECULAR BASIS: IN VITRO STUDIES

In 1973 Goldstein & Brown (17) reported that fibroblasts cultured from the skin of an FH homozygote had a fifty-fold higher level of HMG–CoA reductase activity, the rate-limiting enzyme of cholesterol synthesis, than normal fibroblasts.   Unlike the latter the activity in FH cells was not inhibited by the addition of LDL to the culture medium.   Subsequently they showed that labelled LDL was bound to normal fibroblasts by means of a specific high affinity, saturable mechanism which was lacking in cells from FH homozygotes (18).   In a series of further studies, which they recently summarized (19), these authors postulated that high affinity binding of LDL was dependent upon the presence of specialized receptors on

the cell surface of fibroblasts and that these LDL receptors were
lacking to a greater or lesser extent in FH cells.   Using this
approach they subdivided homozygotes into receptor-negative or
receptor-defective categories (20), whereas cells from heterozygotes
showed only about a 50 % reduction in their complement of LDL receptors.   Under normal circumstances the binding of LDL to these receptors initiates the following chain of events: cellular uptake
of LDL;   incorporation of LDL into lysosomes, with hydrolysis of
both the cholesterol ester and apoB moieties of LDL;   suppression
of HMG-CoA reductase activity and thus of cholesterol synthesis in
microsomes by newly-released free cholesterol, the latter subsequently being re-esterified as a result of the activation of the enzyme
acyl-CoA:cholesterol acyltransferase (19).   In receptor-negative
cells absence of specific binding of LDL precludes any of the subsequent steps from occurring at a normal rate.

An alternative explanation for the defect in FH was proposed
by Stein et al (21).   These authors found no decrease in LDL
binding in cultured fibroblasts from an FH homozygote but showed
a marked decrease in the rate of 'internalization' or cellular uptake of bound LDL.   However, Brown and Goldstein (2) subsequently
claimed to have demonstrated the absence of LDL receptors in the
same strain of cells, and suggested that the discrepancy was due to
methodological differences.

The existence of LDL receptors has also been demonstrated in
cells cultured from the intima and media of human foetal aorta (22)
and in freshly isolated lymphocytes from normal subjects, but not
from patients with homozygous FH;   a 72 hour period of incubation
in an LDL-deficient medium is necessary, however, to unmask these
high affinity binding sites (23).   Fogelman et al (24) had earlier shown that cholesterol synthesis in leucocytes was enhanced by
prior incubation in lipoprotein-deficient serum and that this occurred to a greater extent than normal in cells from FH heterozygotes.
Subsequent studies showed enhanced activity of HMG-CoA reductase
in FH leucocytes, together with evidence of an increased rate of
efflux of cholesterol into the incubation medium (25).   These
authors proposed that leakage of cholesterol from the cell is the
primary defect in FH and that the increased activity of HMG-CoA reductase and consequent enhancement of cholesterol synthesis are
secondary events.

The validity of the LDL receptor hypothesis in the context of
FH has recently been examined by Myant (26).   In his view one of
its main attractions is that it explains the selectivity of the increase in LDL, as opposed to other cholesterol-carrying lipoproteins.
Furthermore the complete or partial absence of LDL receptors is consistent with the most characteristic _in vivo_ abnormality, namely a
reduced fractional rate of catabolism of LDL. The fact that catabolism occurs at all in receptor-negative homozygotes is explicable

on the basis of non-specific uptake and degradation of LDL which
proceeds at similar rates in normal and FH fibroblasts (27), assuming that the latter behave similarly to the cells involved in LDL
catabolism.   However, the receptor hypothesis is less easy to re-
concile with the lack of evidence of any in vivo increase in cholesterol synthesis, since non-specific uptake of LDL into homozygous
FH fibroblasts does not appear to inhibit their HMG-CoA reductase
activity.   But although much remains to be explained nevertheless
the work of Goldstein and Brown has had a considerable impact on
the field of lipoprotein metabolism and has stimulated others to
utilize similar techniques.

MOLECULAR BASIS: IN VIVO STUDIES

Most of the earlier in vivo studies were aimed at measuring
cholesterol turnover in FH.   In most instances there was no evidence of any increase in cholesterol synthesis in FH patients, whether homozygous or heterozygous (26).   An exception was the young
homozygote reported by Bilheimer et al. (28) as having an increased
rate of synthesis, although the absence of any data in control subjects of comparable age throws doubt on the significance of this
observation.   More recently, Bhattacharyya et al. (29) failed to
show any increase in cholesterol synthesis in heterozygotes, in
spite of their being on a nil or low intake of dietary cholesterol.
Furthermore, their absorption of dietary cholesterol has been shown
to be normal (30).   Miettinen and Aro (31) reported that bile acid
excretion was reduced in a family of heterozygotes whom they studied but other workers have found normal values in both homozygotes
and heterozygotes, as well as a normal response to cholestyramine
(32).   Thus it would appear that the absolute rates of synthesis
of both cholesterol and bile acids are usually normal in FH, although there is undoubtedly quite a marked reduction in the fractional rate of clearance of cholesterol from a considerably expanded
rapidly-exchangeable pool (29).

In an attempt to answer the question as to how the rapidly ex-
changeable pool of cholesterol, which includes plasma, gets so en-
larged, Langer et al. (33) measured the rate of turnover of the rad-
ioiodinated protein moiety of LDL, namely apoB, in a group of FH
heterozygotes.   Under steady-state conditions and using the technique of compartmental analysis introduced by Matthews (34), they
found that FH patients differed from control subjects in having a
lower fractional rate of catabolism (FCR) of LDL-apoB, even though
the absolute rate of catabolism, and thus the synthetic rate, was
within normal limits.   Using a similar approach Packard et al.
(35) confirmed the reduced FCR in FH heterozygotes but found that
this was accompanied by a modest increase in LDL synthesis. Reichl
et al. (36) and Simons et al. (37) showed that the FCR of LDL-apoB
was even more reduced in homozygotes, and that this was accompanied

by a marked increase in the rate of apoB synthesis.   Recently,
Thompson and his colleagues (38, 39) performed non-steady-state studies of LDL turnover in FH patients after having reduced the expanded
LDL pool to near-normal size by means of plasma exchange.   The
results showed that the FCR remained remarkably constant, irrespec-
tive of the changing size of the LDL pool, and suggested that the
reduced FCR is an intrinsic feature of the disease rather than a
secondary phenomenon.   This conclusion is strengthened by the
normal FCR of LDL in patients with primary hypobetalipoproteinaemia,
in whom LDL pool size is greatly reduced due to decreased apoB synthesis (40).   Thus it would seem that the hypercholesterolaemia
of FH can be at least partly explained on the basis of defective
catabolism of LDL, the major carrier of cholesterol in plasma.

The mechanism of the increased rate of synthesis of apoB in
homozygotes has also been investigated recently (41).   The results
suggest that more LDL-apoB is secreted into the plasma of these subjects than can be explained on the basis of their normal rate of synthesis and conversion of very low density lipoprotein (VLDL) to LDL.
This is in keeping with the concept of Brown & Goldstein (2) that
the function of the LDL receptor in liver cells is to regulate the
rate of synthesis of apoB, and that absence of such receptors, as
in homozygotes, can lead to the direct secretion of LDL from the
liver into plasma.

The actual site of catabolism of LDL in vivo remains controversial.   The somewhat unphysiological studies of Sniderman et
al. (42) in hepatectomised pigs suggest that LDL catabolism takes
place mainly in extrahepatic tissues, whereas the fact that cholestyramine increases the FCR of LDL as well as enhancing bile acid excretion points to the liver as a major site of catabolism of both
the protein and lipid components of LDL (43).   Experimental data
suggests that hepatic tissues undoubtedly have the ability to bind
LDL, both in vitro (44) and in vivo (45), in an analogous manner to
that seen in fibroblasts, but without quite the same degree of specificity.

Analyses of the amino acid composition of LDL-apoB have consistently failed to show any abnormality in FH (46, 47).   However,
undoubted differences exist between the lipid composition of LDL
from FH patients and normal subjects, the former having a lower proportion of triglyceride than the latter (48).   Of perhaps greater
interest is the increased cholesterol:protein ratio of LDL (49),
which mainly reflects an increase in free cholesterol (50).   This
is accompanied by a decrease in the ratio of lecithin:sphingomyelin
(51, 52).   These abnormalities are more marked in homozygotes
than in heterozygotes but become less abnormal following plasma exchange (53).   This suggests that the increased content of sphingomyelin and cholesterol are causally related and that they are probably secondary to the prolonged half-life of LDL in the circulation.

In this context it is of interest that fibroblasts from FH homozygotes also have an increased content of sphingomyelin, which diminishes after they have been incubated in a lipoprotein-free medium (54).   It seems unlikely that the abnormalities of LDL composition are primary phenomena since analogous changes have been observed in animals made hyperbetalipoproteinaemic by methods which decrease LDL catabolism (51, 55).

MODEL FOR REGRESSION OF ATHEROSCLEROSIS

The premature onset of atherosclerosis is an inevitable accompaniment of FH.   This tendency is particularly marked in homozygotes, many of whom die suddenly in childhood or in their teens (5). Few, if any, have survived beyong the age of 30 (1).   The considerable risks of developing coronary and peripheral vascular disease in heterozygotes are also well documented, and the averaged results of three surveys are shown in the Table.   It is clear that males are affected at an earlier age than females, although this sex advantage is apparently lost in females who smoke cigarettes (14). Slack (56) found that male heterozygotes have a 23 % expectation of death from coronary heart disease before the age of 50.   Heiberg (15, 57) studied several Norwegian families with FH and found that half the males were dead by the age of 66, whereas half the females could expect to live to the age of 75.   In his series both the age of onset of ischaemic symptoms and the appearance of xanthomata were related to the height of the serum cholesterol. Few patients with serum cholesterol levels below 10 mmol/l developed xanthomata. But half the males and half the females who did develop xanthomata were dead within 8 to 12 years respectively of their first appearance.

TABLE

Risks of Coronary Heart Disease in FH.   Pooled data from Slack (56), Stone et al (65) and Beaumont et al (14)*

Age	Males	Females
30	5 %	0 %
40	29 %	20 %
50	60 %	31 %
60	69 %	45 %

* Reference (14) included patients with Peripheral Vascular Disease.

During the past 10 years considerable efforts have been made to reduce the premature morbidity and mortality from coronary heart disease in FH by reducing the severity of the hypercholesterolaemia. The mainstay of treatment has been the anion-exchange resin cholestyramine which specifically lowers LDL levels. At the present time a large-scale trial of the efficacy of this drug in the primary prevention of coronary heart disease in patients with type II hyperlipoproteinaemia (some of whom have FH) is being carried out in the United States, but the results will not be available for sometime. Another proposed trial, but of secondary intervention, has recently been described by Buchwald et al (58). This is based on the efficacy of ileal-bypass as a means of lowering the serum cholesterol and involves the use of coronary arteriography as a means of assessing progression or regression of established coronary heart disease in patients with hypercholesterolaemia, some of whom will presumably have FH.

Ileal bypass is undoubtedly as effective as cholestyramine in the management of heterozygous FH but neither form of treatment adequately controls the hypercholesterolaemia of homozygotes. An alternative approach was conceived by Starzl et al (59) who performed a portacaval shunt in a young female homozygote; the ensuing reduction in hypercholesterolaemia was said to be accompanied by evidence of regression of her atheroma (60). Subsequent experience with this operation has been less encouraging, however, since on average it results in only an approximately 30 % reduction in serum cholesterol levels (61). The mechanism whereby portacaval shunts influence lipoprotein metabolism remains controversial but a decrease in hepatic cholesterol synthesis has been documented both in animals (62) and in a patient with homozygous FH (28).

Because of the difficulty of controlling the hypercholesterolaemia of homozygous and severe heterozygous FH by conventional means, Thompson et al (63) introduced the use of plasma exchange. This is performed with a continuous flow blood cell separator (Aminco "Celltrifuge"), as illustrated in Fig.1. Heparinised venous blood from the patient is continuously separated into plasma and cellular components by centrifugation. By means of an ingenious seal and peristaltic pumps the plasma is collected for disposal, whereas the cellular components are returned to the patient via a vein in the opposite arm, after being reconstituted with plasma protein fraction (PPF). The latter is a cholesterol-free solution containing mainly human serum albumin. In general 3-4 litres of plasma are exchanged on each occasion at whole blood flow rates of up to 70 ml per minute, the entire procedure lasting about 2 hours.

The rationale for plasma exchange is that the hypercholesterolaemia of FH is largely due to defective catabolism of LDL, as illustrated by Fig.2. Thus physical removal of LDL, which is mainly located intravascularly (32), is a logical approach to the

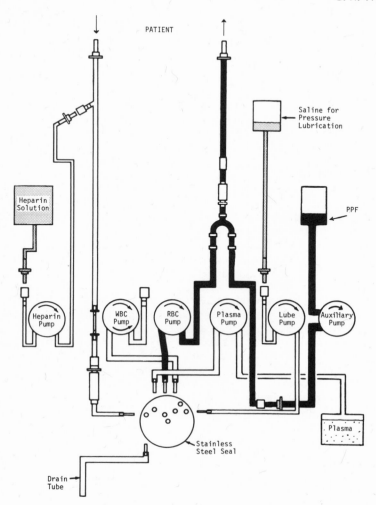

Fig.1. Simplified circuit diagram of "Celltrifuge". The stainless steel seal on top of the centrifuge bowl is continuously lubricated with sterile saline.

problem.   Immediately after plasma exchange the serum cholesterol level decreases to one-third of the pre-exchange value, and then gradually rises at the rate of about 5 mmol/l per week in homozygotes and 2.5 mmol/l per week in heterozygotes, the rate of rise being slowed to some extent by hypocholesterolaemic agents.   However, even so, effective control of hypercholesterolaemia necessitates weekly plasma exchange in all homozygotes and also in some heterozygotes, although in others plasma exchange every 2 weeks in probably sufficient.

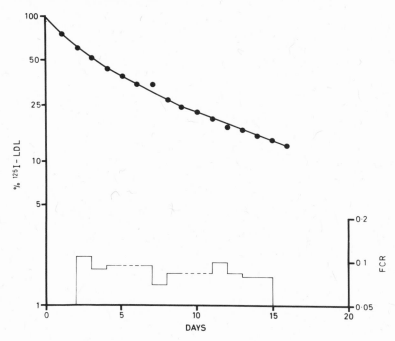

Fig.2. Slow turnover of ^{125}I-LDL in patient with homozygous FH ,
manifested by the prolonged half-life of LDL in plasma (7.4 days)
and reduction of the daily fractional catabolic rate (FCR) to less
than 0.1.

Each patient selected for a course of plasma exchange at the
Hammersmith Hospital first undergoes coronary angiography and
aortography, in order to delineate the extent of atheroma in these
vessels.   These investigations are repeated after a period of
approximately 2 years of plasma exchange, so as to obtain an objective assessment of each patient's progress.   Conventional drug
therapy with β-blockers for angina and hypocholesterolaemic agents
(usually cholestyramine plus either nicotinic acid or clofibrate) is
maintained throughout.   To date 4 homozygotes and 3 heterozygotes
have been or are regularly undergoing plasma exchange. Two patients
have died; one of them, a homozygous female, died following an aorto-
coronary bypass, the other, a male heterozygote, from a myocardial
infarct.   Neither death was attributable to plasma exchange. Of
the other five patients two are male homozygotes who remain asympto-
matic and in whom cutaneous and tendon xanthomata show signs of resolution.   Plasma exchange was discontinued in the fourth homozygote because the smallness of her veins precluded good flow rates;
subsequently she underwent a porta-caval shunt.   Two male
heterozygotes remain well and have had some amelioration of their
angina.   In one of them weekly plasma exchange maintains serum

cholesterol levels consistently within the normal range (< 6.5 mmol/l).

There have been no untoward reactions during the last 100 plasma exchanges, in particular there has been no evidence of sensitisation to PPF, no haemorrhage or sepsis and no cardiovascular complications.   All patients have remained Australia antigen negative. Episodes of hypocalcaemia and hypokalaemia have been prevented by the routine addition of 1.5 mmol potassium (as potassium chloride) and 0.45 mmol calcium (as calcium gluconate) to each 400 ml of PPF. Using this approach serum electrolytes remain normal throughout. The platelet count decreases by up to 50 % by the end of the procedure but returns to normal within one week.   Thus plasma exchange at weekly or fortnightly intervals has proved to be a safe and practicable procedure which results in substantially better control of hypercholesterolaemia than could otherwise be achieved. Whether this will result in demonstrable improvement of coronary atheroma remains to be seen.   However, the recent report of Barndt et al (64) is encouraging, in that it showed angiographic evidence of regression of femoral atheroma in FH heterozygotes in whom adequate control of hyperlipoproteinaemia had been achieved by conventional means.

REFERENCES

1. Fredrickson, D.S.,and R.I. Levy, in The Metabolic Basis of Inherited Disease, Third Edition.   (Eds)   J.B. Stanbury, J.B. Wyngaarden and D.S. Fredrickson, McGraw Hill, New York (1972) pp. 545–614.

2. Brown, M.S., and J.L. Goldstein.   New Engl. J. Med. 294: 1386 (1976).

3. Armstrong, M.L., E.D. Warner, and W.E. Connor.   Circulation Res. XXVII:59 (1970).

4. Daoud, A., J. Jarmolych, J.M. Augustyn, K.E. Fritz, J.K. Singh, and K.T. Lee.   Arch. Pathol. Lab. Med. 100:372 (1976).

5. Cook, C.D., H.L. Smith, C.W. Geissen, and G.L. Berdez. Amer. J. Dis. Child. 73:326 (1947).

6. Rigadon, R.H., and G. Willeford.   J. Amer. Med. Ass. 142: 1268 (1950).

7. Stanley, P., C. Chartrand, and A. Davignon.   New Engl. J. Med. 273:1378 (1965).

8. Bialostozky, D., M.Luengo, C. Magos, and E. Zorrilla.   Amer. J. Cardiol. 36:509 (1975).

9. Barr, D.P., S. Rothbard, and H.A. Eder.   J. Amer. Med. Ass. 156:943 (1954).

10. Wennevold, A., and J.G. Jacobsen,   Amer. J. Med. 50:823 (1971).

11. Roberts, W.C., V.J. Ferrans, R.I. Levy, and D.S. Fredrickson. Amer. J. Cardiol. 31:557 (1973).

12. de Gennes, J.L., J. Rouffy, and F. Chain.   Bull. Soc. Med. Hop. Paris 119:569 (1968).

13. Kwiterovich, P.O., D.S. Fredrickson, and R.I. Levy.   J. Clin. Invest. 53: 1237 (1974).

14. Beaumont, V., B. Jacotot, and J.-L. Beaumont.   Atherosclerosis 24:441 (1976).

15. Heiberg, A.   Acta med. Scand. 198:249 (1975).

16. Bloch, A., R.E. Dinsmore, and R.S. Lees.   Lancet i:928 (1976).

17. Goldstein, J.L., and M.S. Brown.   Proc. Nat. Acad. Sci. U.S. 70:2804 (1973).

18. Brown, M.S., and J.L. Goldstein.   Proc. Nat. Acad. Sci. U.S. 71:788 (1974).

19. Brown, M.S., and J.L. Goldstein.   Science 191:150 (1976).

20. Goldstein, J.L., S.E. Dana, G.Y. Brunschede, and M.S. Brown. Proc. Nat. Acad. Sci. U.S. 72:1092 (1975).

21. Stein, O., D.B. Weinstein, Y. Stein, and D. Steinberg. Proc. Nat. Acad. Sci. U.S. 73:14 (1976).

22. Goldstein, J.L., and M.S. Brown.   Arch. Pathol. 99:181 (1975).

23. Ho, Y.K., M.S. Brown, D.W. Bilheimer, and J.L. Goldstein. J. Clin. Invest. 58:1465 (1976).

24. Fogelman, A.M., J. Edmond, A. Polito, and G. Popjak.   J. Biol. Chem. 248:6928 (1973).

25. Fogelman, A.M., J. Edmond, J. Seager, and G. Popjak.   J. Biol. Chem. 250:2045 (1975).

26. Myant, N.B., in Cholesterol Metabolism and Lipolytic Enzymes. (Ed.) J. Polonovski, Masson Publishing U.S.A., New York (1977) pp. 39-52.

27. Goldstein, J.L., and M.S. Brown.   J. Biol. Chem. 249:5153 (1974).

28. Bilheimer, D.W., J.L. Goldstein, S.M. Grundy, and M.S. Brown. J. Clin. Invest. 56:1420 (1975).

29. Bhattacharyya, A.K., W.E. Connor, and A.A. Spector.   J. Lab. Clin. Med. 88:202 (1976).

30. Connor, W.E., and D.S. Lin. J. Clin. Invest. 53:1062 (1974).

31. Miettinen, T.A., and A. Aro. Scand. J. Clin. Lab. Invest. 30:85 (1972).

32. Moutafis, C.D., L.A. Simons, N.B. Myant, P.W. Adams, and V. Wynn.   Atherosclerosis 26:329 (1977).

33. Langer, T., W. Strober, and R.I. Levy.   J. Clin Invest. 51: 1528 (1972).

34. Matthews, C.M.E.   Phys. in Med. Biol. 2:36 (1957).

35. Packard, C.J., J.L.H.C. Third, J. Shepherd, A.R. Lorimer, H.G. Morgan, and T.D.V. Lawrie. Metabolism 25:995 (1976).

36. Reichl, D., L.A. Simons, and N.B. Myant. Clin. Sci. Molec. Med. 47:635 (1974).

37. Simons, LA., D. Reichl, N.B. Myant, and M. Mancini. Atherosclerosis 21:283 (1975).

38. Thompson, G.R., and N.B. Myant. Atherosclerosis 23:371 (1976).

39. Thompson, G.R., T. Spinks, A. Ranicar, and N.B. Myant.
    Clin. Sci. Molec. Med. 52:361 (1977).
40. Sigurdsson, G., A. Nicoll, and B. Lewis.   Metabolism 26:25
    (1977).
41. Thompson, G.R., A.K. Soutar, and N.B. Myant.   Circulation
    54, Suppl. II:26 (1976).
42. Sniderman, A.D., T.E. Carew, J.G. Chandler, and D. Steinberg.
    Science 183:526 (1974).
43. Levy, R.I., and T. Langer. Advanc. Exp. Med. Biol. 21:7
    (1972).
44. Bachorik, P.S., J.N. Livingston, J. Cooke, and P.O. Kwiter-
    ovich.   Biochem. Biophys. Res. Commun. 69:927 (1976).
45. Anderson, J.M., F.O. Nervi, and J.M. Dietschy.   Biochim.
    Biophys. Acta 486:298 (1977).
46. Pinon, J.C., and P. Laudat.   Biochim. Biophys. Acta 187:
    144 (1969).
47. Gotto, A.M., W.V. Brown, R.I. Levy, M.E. Birnbaumer, and
    D.S. Fredrickson.   J. Clin. Invest. 51:1486 (1972).
48. Slack, J., and G.L. Mills. Clin. Chim. Acta 29:15 (1970).
49. Beaumont, J.-L., B. Swynghedauw, and V. Beaumont.   Rev.
    Franc. Etud. Clin. Biol. 10:221 (1965).
50. Shattil, S.J., J.S. Bennet. R.W. Colman, and R.A. Cooper.
    J. Lab. Clin. Med. 89:341 (1977).
51. Howard, A.N., V. Blaton, D. Vandamme, N. Van Landschoot, and
    H. Peeters.   Atherosclerosis 16:257 (1972).
52. Mills, G.L., C.E. Taylaur, and M.J. Chapman.   Clin. Sci.
    Molec. Med. 51:221 (1976).
53. Jadhav, A.V., and G.R. Thompson.   Clin. Sci. Molec. Med.
    52:18 (1977).
54. Chatterjee, S., C.S. Sekerke, and P.O. Kwitervich.   Proc.
    Nat. Acad. Sci. U.S. 73:4339 (1976).
55. Klauda, H.C., and D.B. Zilversmit.   J. Lipid Res. 15:593.
    (1974).
56. Slack, J.   Lancet ii:1380 (1969).
57. Heiberg, A.   Clin. Genet. 9:92 (1976).
58. Buchwald, H., R.B. Moore, and R.L. Varco.   Lipids 12:53
    (1977).
59. Starzl, T.E., H.P. Chase, C.W. Putnam, and K.A. Porter.
    Lancet ii:940 (1973).
60. Starzl, T.E., H.P. Chase, C.W. Putnam, and J.J. Nora.
    Lancet ii:714 (1974).
61. Russell, D., C. Mieny, K.W. Heimann, M. Dinner, S.E. Levin,
    B.I. Joffe, D. Mendelsohn, A. Mega de Andrade, S. Lieber-
    thal, and H.C. Seftel.   Lancet ii:1205 (1976).
62. Chase, H.P., and T. Morris.   Atherosclerosis 24:141 (1976)
63. Thompson, G.R., R. Lowenthal, and N.B. Myant.   Lancet i:
    1208 (1975).
64. Barndt, R., D.H. Blankenhorn, D.W. Crawford, and S.H. Brooks.
    Ann. Intern. Med. 86:139 (1977).
65. Stone, N.J., R.I. Levy, D.S. Fredrickson, and J. Verter.
    Circulation 49:476 (1974).

BIOCHEMISTRY OF PLASMALIPOPROTEINS IN LIVER DISEASES

D.SEIDEL

Medizinische Universitätsklinik

Bergheimerstrasse 58, D-6900 HEIDELBERG, West Germany

A LIPOPROTEIN-X

The liver is the major site of synthesis of plasmalipoproteins and has central functions in lipoprotein catabolism. Since abnormal serum lipid concentrations are often associated with abnormal liver function it is reasonable to anticipate that liver disease will result in alterations of plasmalipoproteins, the protein lipid particle in which form all serum lipids circulate. Undoubtedly, the example best known to clinicians is the hypercholesterolemia and hyperphospholipidemia accompanying biliary obstruction or cholestasis, a phenomenon recorded more than a century ago by FLINT (1). The increased serum cholesterol in such patients is in form unesterified cholesterol, thus, the percentage of serum cholesterol present in esterified form is decreased, although in absolute amounts esterified cholesterol concentrations usually remain normal as long as liver function otherwise remained undisturbed, or as long as the plasma activity or the lecithin:cholesterol:acyltransferase (LCAT enzyme) is normal (2).

In recent years studies from different laboratories have well documented that the characteristic plasma lipid values found in cholestasis are preliminary due to the presence of an abnormal plasmalipoprotein, which is normally found within the LDL-density class (3, 4) and which may be isolated in intact form by combination of various physicochemical techniques (5, 6). This compound has been designed lipoprotein-X.

Lipoprotein-X (LP-X) seems to have the characteristics of a vesicle with a mean diameter of 600 Å with a tendency to aggregate as judge from negatively stained electronmicrographs at pH 7 and room temperature (7, 10).

The protein lipid composition of isolated LP-X is unique : 6%
protein, 66% phospholipids, 22% unesterified cholesterol, 3% cholesterolester and 3 % triglycerides (5). The amount of bile acids bound
to LP-X may very greatly causing pronounced changes in the physico-
chemical properties particularly in the hydrated density of the particle. The aspect has only very recently been considered. The protein portion of LP-X consists of 40% albumin, located in the core of
the particle or masked by lipids and of apo-C and apo-D which are
considered to be surface proteins of LP-X (6). It has been proposed
that the specific combination of albumin with Apo-C peptides and
Apo-D plays an important role in maintaining the structural integrity of this lipoprotein particle. In addition to the apoproteins and
to albumin recent data from our laboratory demonstrate that major
portions of plasma GGT are also bound to LP-X.

One important characteristic behaviour of LP-X is its electrophoretic mobility towards the cathode on agar-gel which is in con-
traxt to all normal plasmalipoproteins migrating to the anode. Because of its simplicity, high sensitivity and specificity to inden-
tify LP-X, this feature has widely been used for LP-X determination
and measurement.

Since the introduction of this method into clinical chemistry
some years ago (12) a large body of literature uniformly documents
the significance of this test in the differential diagnosis of jaundice. It separates with the highest specificity and sensitivity a-
mong all blood chemical papreters cholestasis from noncholestasis
(13, 21).

It has previously been demonstrated that LP-X appears in appre-
ciable amounts in dog and rat plasma within the first 24 hrs following surgical ligation of the common bile duct. The same time seems
to be necessary for LP-X to appear in patients with estrahepatic biliary obstruction. It was also demonstrated that during extrahepatic biliary obstruction bile passes from the bile ducts to the hepatic lymphatics, from there to the thoracic lymphatic duct and finally into the blood stream. It is alo well established that liver ex-
creeds lipids, predominantly phospholipids and unesterified cholesterol with the bile fluid, thus the lipid composition of LP-X isolated from serum shows great similarity to the lipids found in bile.
Even the phosphatide distribution and phosphatide fatty acids of
bile are almost identical to that of LP-X (7). These facts taken
together strongly suggested a close relationship of bile lipids and
LP-X.

Human bile, when subitted to agar or agarose electrophoresis
reveals a lipoprotein ban migrating towards the anode, which can be
visualized by polyanionprecipitation, as well as with a lipid stain
or protein stain (11). Immunochemically albumin and immunoglobulins
as well as other plasmaproteins can be identified but no immunopre-
cipitation reaction is obtained with antibodies to any of the major
apolipoproteins Apo-A, Apo-B, Apo-C, apo-D and apo-E. Under the

electronmicroscope it is not possible to detect reproducable any particular structures occasionally some lamellar structures are apparent. LP-X, however, with its characteristic structure under the electronmicroscope or even more specifically with its typical migration towards the cathode on agar electrophoresis cannot be detected in neither native nor water dialyzed bile. All bile lipids may be separated from bile in form of an albumin-lipid complex (designated bile LP). This complex shows a lipid composition very similar to LP-X but displays different physico-chemical characteristics. When bile, however, is incubated in vitro with either pure albumin or with total (LP-X negative) serum, LP-X is formed and may be isolated showing an electrophoretic behaviour and other physochemical characteristics similar to that found in the serum of patients suffering from cholestatic liver disease (11). Addition of various amounts of isolated VLDL, LDL or HDL as well as purified immunoglobulins to native bile does not result in LP-X formation.

The amount of LP-X formed from bile after addition of albumin or serum depends to a high degree on the concentration of native bile, in particular on the concentration of bile lipoprotein, bile salts and the amount of albumin added tot he system. Comparison of LP-X formed by albumin alone (designated LP-X alb) or by serum (LP-X serum) with native LP-X revealed a very similar lipid composition. Significant differences are apparent, however, in their protein moiety and the structure properties as judged by electronmicroscopy and electrophoresis on agar-gel. LP-X alb shows somewhat larger and more heterogenous particles on electronmicrographs, it migrates further to the cathode on agar electrophoresis as does LP-X and is devoid of Apo-C and Apo-D. After in vitro incubation with HDL or VLDL this material as well as LP-X serum shows these apolipoproteins, the latter is almost indistinguishable from native LP-X. From these in vitro experiments which are strongly supported by in vivo studies in dog it seems evident that LP-X formation takes place when bile LP is converted into LP-X by the action of albumin in the proper ratio to bile salts. Thus albumin, rather than other Apo-X peptides seems to be the key protein in LP-X formation and in maintaining the unique physico-chemical characteristics of LP-X, although Apo-C and Apo-D in addition may have some stabilizing effect on the structure of this abnormal plasma lipoprotein.

The plasma influx of biliary phospholipids and free cholesterol in form of bile lipoprotein or LP-X respectively may be of help for our understanding of cholesterol metabolism and the mechanism of hypercholesterolemia in cholestasis. It is well established that cholesterol synthesis but not so phospholipids synthesis is increased in cholestasis. Moreover, phospholipids in a high ratio to cholesterol as in LP-X stimulate hepatic cholesterol synthesis and recent results from our laboratory show that LP-X, probably because it is devoid of Apo-B, is not able to inhibit hepatic cholesterol synthesis in contrast to Apo-B carrying lipoproteins (22).

The role of lecithin:cholesterol:acyltransferase, the LCAT-
enzyme in the metabolism of plasmalipoproteins is still a matter of
speculation. This holds particularly for the cholestatic conditions.
LCAT is likely to be produced in the liver and the lipoprotein alterations in liver disease are in many respects similar to those found
in familial LCAT-deficiency. Although some experimental data are
available (23) on the relationship of LCAT and lipoproteins in liver
disease, our knowledge in this field is still very lipited. The primary reason for this seems to be that the exact mode of action of
this enzyme, which is responsible for esterification of lipoprotein
cholesterol in plasma is still not clear. It was believed that HDL
or lipids in HDL are the major cholesterol and lecithin substrate
for the enzyme and that Apo-A$_1$ was the primary cofactor. However,
there are also reports indicating that VLDL rather than HDL should
be considered as substrate and that Apo-D or even a protein in the
d 1.25 g/ml fraction are the most potent activators for the enzyme.
It is without doubt that cholesterol, lecithin and a protein activator is required for the LCAT reaction but more than that the structural arrangements of these moieties within the lipoprotein particle
must carefully be considered. Evidence for the great importance of
this aspect is provided by studies of interactions between membrane
lipids and lipolytic enzymes. The relationship, if there is any,
between LP-X and LCAT in cholestasis is not clear. LCAT-activity
may be normal elevated or decreased in cholestasis despite the presence of LP-X. According to the carefully performed studies by
RITLABD and GJONE (24) an inverse correlation between LCAT-activity
holds only for the concentration of plasma cholesterolester concentration but not for LP-X concentration in neither cholestatic liver
disease nor in familial LCAT deficiency. The fact that LP-X is often the only plasmalipoprotein abnormality particularly in forms of
mild cholestasis accompanied with an elevated LCAT-activity and normal ester ratios in the other lipoproteins may indicate that LP-X
is not a particularly good substrate for the LCAT enzyme. Our own
in vitro studies in which isolated LP-X was incubated with a partially purified LCAT fraction supports this idea and are in agreement
with data by MAGNANI (25). It may be speculated that the mechanism
of LP-X formation in familial type of LCAT deficiency is entirely
different from that in cholestasis. It may, however, also be a result of the particular physicochemical conditions of the lipids in
this disease.

B   HYPERTRIGLYCERIDEMIA IN LIVER DISEASE

Hypertriglyceridemia as a consequence of alcohol intake is a
well known phenomenon (26) which, however, is not fully understood
in the pathophysiological mechanisms.

Only recently particular attention has been paid to the hypertriglyceridemia, which may be found in some patients with liver dys-
function (27, 31), a metabolic situation most frequently accompanied
by severe cholestasis. It was demonstrated that the majority of

triglycerides of such patients is found within the LDL lipoprotein
fraction. The isolation of a large (300-700 Å in diameter) triglyceride rich low density lipoprotein (d 1.019 - 1.063 g/ml) different
from LP-X and normal beta-lipoproteins was later achieved by a combination of ultracentrifugation, cold ethanol fractionation and affinity chromatography (31). This lipoprotein has been designated
beta$_2$-lipoprotein because it shares certain physicochemical characteristics and chemical properties with normal beta-lipoproteins.
Detailed analyses of the lipoprotein families of such patients indicated that the increased concentration of LDL triglycerides as well
as the decreased VLDL-TG/ 1.006 g/ml-TG ratio and hereby the hypertriglyceridemia of such patients with liver disease is primarily
due to the presence of beta$_2$-LP. Since the protein moiety of this
low density lipoprotein consists of Apo-B and Apo-C and possibly
also of trace amounts of Apo-A it is likely that beta$_2$-LP derives
from lipoproteins of lower density which are known to contain all
major apoproteins. This possibility is supported by a dietary experiment (31). The decrease of its plasma concentration on a low
fat diet may indicate its nature as a degradation product or intermediate particle of chylomicron metabolism, although an additional
relationship to VLDL catabolism cannot be ruled out with our present
knowledge. One significant mechanism responsible for the accumulation of this low density lipoprotein is the markedly reduced hepatic lipase activity, which we first demonstrated in such patients
(31) and which may be relevant for the mechanism of their hypertriglyceridemia.

C   CHANGES OF THE VLDL AND HDL FRACTION OCCURING IN LIVER DISEASE

     Besides striking abnormalities of the LDL-fraction changes from
normal can also be found in other density classes (32). Using the
analytical ultracentrifuge a diminuition of the HDL-fraction was
demonstrated. The lipoprotein electrophoresis is frequently altered showing only one broad beta-band in beta-position, the pre-beta-
and alpha-bands are lacking (32).

     The absence of pre-beta- and alpha-lipoprotein bands on lipo-
protein electropherograms does, however, not necessarily allow the
conclusion that the corresponding lipoprotein density fractions are
absent in the plasma sample.

     The VLDL-fraction may even be increased in such patients. Iso-
lated VLDL exhibit not pre-beta- but beta-mobility on agarose-elec-
trophoresis and show a marked decrease of Apo-C in the apoprotein
moiety. Incubation of VLDL isoloated from liver diseased patients
with normal HDL restores the normal pre-beta-mobility, indicating
a transfer most likely of apoproteins from HDL to VLDL.

     The lacking alpha-lipoprotein band is usually accompanied by a
decrease of the HDL-concentration to about 75% on the average. The
plasma concentration of A-I and A-II together, however, is not found

to be decreased that much. The immunoelectrophoretical pattern of
HDL from liver diseased patients against alpha-lipoproteins strongly suggests a dissociation of apo-A-I and apo-A-II forming two non
identical protein molecules and corresponding lipoproteiñ families.
Such HDL fractions hardly contain any cholesterol or triglycerides
but phospholipids.

In addition to the altered alpha-lipoproteins a second abnormal
lipoprotein has been isolated from the HDL-fraction of patients suffering from severe prolonged cholestasis. It was designated LP-E,
showing primarly if not only the arginine-rich apoprotein (34). On
electonmicrographs it exhibits rouleaux formations like LP-X and
migrates in beta-position on agarose electrophoresis. In contrast
to LP-X it migrates also on agar towards the anode.

Since A-I and A-II are impaired to bind neutral lipids in dissociated form they are detectable on lipoprotein electrophoresis. The
lack of neutral lipids may also cause the deficiency of HDL in LP-C
and may, in addition, be of importance with regard to the regulation
of chylomicron and VLDL catabolism. To clarify the nature of the
structural abnormality of high density lipoproteins in liver disease
much further work needs to be done. The concentration of A-I and
A-II has to be determined in the plasma of these patients and under
the course of the disease. Data of such studies should be related
to other parameters of lipid and lipoprotein metabolism.

REFERENCES

1.   FLINT, A. jr. : Experimental researches into a new excretory
     function of the liver, consisting in the removal of choleste-
     rine from the blood, and its discharge from the body in the
     form of stercorine.
     Amer. J. med. Sci. 44, 305 (1962).

2.   SEIDEL, D. : Hyperlipoproteinaemia bei Erkrankungen der Leber.
     Schweiz. med. Wschr. 105, 857 (1975).

3.   RUSS, E.M., RAYMUNT, J. and D.P.BARR : Lipoproteins in primary
     biliary cirrhosis.
     J. Clin. Invest. 35, 133 (1956).

4. SWITZER, S. : Plasma lipoproteins in liver disease. I. Immuno-
    logically distinct low-density lipoproteins in patients with bi-
    liary obstruction.
    J. Clin. Invest. 46, 1855 (1967).

5. SEIDEL, D. ALAUPOVIC, P. and R.H. FURMAN : A lipoprotein charac-
    terizing obstructive jaundice. I. Method for quantitative sepa-
    ration and identification of lipoproteins in jaundiced subjects.
    J. Clin. Invest. 48, 1211 (1969).

6. SEIDEL, D. ALAUPOVIC, P., R.H. FURMAN and W.J. McCONATHY : A lipoprotein characterizing obstructive jaundice. II. Isolation and
   partial characterization of the protein moieties of low density
   lipoproteins.
   J. Clin. Invest. 49, 2396 (1970).

7. PICARD, J. and D. VEISSIERE : Séparation des lipoprotéines séri-
   ques anormales dans la cholestase.
   C.R. Acad. Sci. (Paris) Ser. D. 270, 1845 (1970).

8. MILLS, G.L., SEIDEL, D. and P.ALAUPOVIC : Ultracentrifugal characterization of a lipoprotein occuring in obstructive jaundice.
   Clin. Chim. Acta 26, 239 (1969).

9. HAMILTON, R.L., HAVEL, R.J. KANE, J.P., BLAUROCK, E.A. and
   T.SATA : Cholestasis : lamellar structure of the abnormal human
   serum lipoprotein.
   Science (Wash. D.C.) 172, 475 (1971).

10.SEIDEL, D., AGOSTINI, B. and P.MUELLER : Structure of an abnormal plasma lipoprotein (LP-X) characterizing obstructive jaundice.
   Biochim. Biophys. Acta 260, 146 (1972).

11.NANZATO, E., FELLIN, R., BAGGIO, G., WALCH, S., NEUBECK, W. and
   D. SEIDEL : Formation of lipoprotein-X. Its relationship to
   bile compounds.
   J. Clin. Invest. 57, 1248 (1976).

12.NEUBECK, W. and D. SEIDEL : Direct method for measuring lipoprotein-X in serum.
   Clin. Chem. 21, 853 (1975).

13.SEIDEL, D., SCHMITT, E.A. and P. ALAUPOVIC : Ein abnormes
   Low-density-Lipoprotein bei Cholestase. II. Bedeutung in der
   Differentialdiagnose des Ikterus.
   Dtsch. Med. Wochenschr. 95, 1805 (1970).

14.POLEY, J.R., ALAUPOVIC, P., McCONATHY, W.J., SEIDEL, D.,
   ROY, C.C. and A. WEBER : Diagnosis of extrahepatic biliary obstruction in infants by immunochemical detection of LP-X and
   modified 133 I-Rose Bengal excretion.
   J. Lab. Clin. Med. 81, 325 (1973).

15.SEIDEL, D., GRETZ, H. and C. RUPPERT : Significance of the LP-X
   test in differential diagnosis of jaundice.
   Clin. Chem. 19, 86 (1973).

16.PREXL, H-J. and W. PETEK : Die Bedeutung des Lipoprotein-X und
   der Serumcholestase in der präoperativen Diagnostik des Ver-
   schlussikterus.
   Chirurg. 44, 310 (1973).

17.RITLAND, S., BLOMHOFF, J.P., ELGJO, K. and E. CJONE : Lipoprotein-X (LP-X) in liver disease.

18. VERGANI, C., PIETROGRANDE, M., GRONDONA, M.C. and M. PIZZOLATO :
    Study of the abnormal lipoprotein-X in obstructive and non-ob-
    structive jaundice.
    Clin. Chim. Acta 48, 243 (1973).

19. ROSTI, D. and C.G. DE GASPARI : Valoré diagnostico della LP-X
    negli itteri neonatali.
    Minerva Pediatr. 26, 582 (1974).

20. MAYER, K. : Der Wert der LP-X-Bestimmung zur Erfassung einer
    Cholestase.
    In : Jahrestagung der Deutschen, Österreichischen und Schweize-
    richen Gesellschaft für klinische Chemie, Freiburg, West Germany,
    März 16-13 (Abstr.) 1975).

21. FISCHER, M., FALKENSAMMER, C., BARONACH, G., WAKETUCH, S.,
    KRONBERGER, O. and H. SCHNACK : Zur Diagnose der Cholestase :
    Lipoprotein-X (LP-X).
    Wien. Klin. Wochenschrift 87, 524 (1975).

22. LIERSCH, M., HEUCK, C.C., BAGGIO, G. and D. SEIDEL : Regulation
    of Cholesterol Biosynthesis in Rat Liver by Two Different Plas-
    ma Lipoproteins.
    Atherosclerosis 26, 505 (1977).

23. McINTYRE, N., HARRY, D.S. and A.I.G. PEARSON : The Hypercholes-
    terolaemia of Obstructive Jaundice.
    Gut 16, 379 (1975).

24. RITLAND, S., BLOMHOFF, J.P. and E. GJONE : Lecithin:Cholesterol
    Acyltransferase and Lipoprotein-X in Liver Disease.
    Clin. Chim. Acta 49, 251 (1973).

25. MAGNANI, H.N. : The influence of LP-X and other lipoproteins as-
    sociated with hepatic dysfunction on the activity of lecithin:
    cholesterol acyltransferase.
    Biochem. Biophys. Acta 450, 390 (1976).

26. BARAONA, E. and LIEBER, L.S. : Effects of chronic ethanol fee-
    ding on serum lipoprotein metabolism in the rat.
    J. Clin. Invest. 49, 769 (1970).

27. KLöR, U., DITSCHUNEIT, H.H., RABOW, D. and H.DITSCHNUNEIT :
    Further characterization of dyslipoproteinemia in hepatic disease.
    Abstr. Eur. J. Clin. Invest. 2, 291 (1972).

28. ALCINDOR, L.G., INFANTE, R. and J.CAROLI : Plasma VLDL catabo-
    lism in cholestasis. Abstr. 5th meeting of the int. Ass. for
    the study of the liver. Versailles, July 1972.

29. PEARSON, A.J.G. : Triglycerides in obstructive liver disease.
    Abstr. 5th meeting of the int. Ass. for the study of the liver.
    Versailles, July 1972.

30. FELLIN, R. and D. SEIDEL : Behaviour of serum lipoproteins in
    cholestasis.

   1. Int. Symp. on cholestasis, Florence, June 1973.
   Proc. of the Symposium.

31.MüLLER, P., FELLIN, R., LAMBRECHT, J., AGOSTINI, B., WIELAND,
   H., ROST, W. and D. SEIDEL : Hypertriglyceridemia Secondary to
   Liver Disease.
   Europ. J. Clin. Invest. 4, 419 (1974).

32.SEIDEL, D. GRETEN, H., GEISEN, H.P., WENGELER, H. and H.WIELAND :
   Further aspects in the characterization of high and very low
   density lipoproteins in patients with liver disease.
   Europ. J. Clin. Invest. 2, 359 (1972).

33.GOFMAN, J. : The serum lipoprotein transport system in health,
   metabolic disorders, atherosclerosis and coronary artery
   diseases.
   Plasma 2, 484 (1954).

34.SEIDEL, D. et al. : unpublished observations.

THE ROLE OF LIPOPROTEINS IN HUMAN ATHEROSCLEROSIS (Lecture I)

K.W. Walton

Department of Experimental Pathology
University of Birmingham
Birmingham 15   England

The atherosclerotic plaque at all stages of its development, and wherever it occurs in the arterial tree, is characterised by lipid deposition. Earlier suggestions that the lipid is derived solely from breakdown ("fatty metamorphosis") of intrinsic components of the arterial wall (1), or originates from blood clots or cellular components of the blood deposited on the endothelial surface and later incorporated into the wall (2), are no longer widely entertained (for review, see Ref.3). Instead, current research seems broadly to support the hypothesis of Virchow (4) that the lipid in atherosclerotic lesions derives from the "insudation" of plasma into the arterial wall. However the original hypothesis requires modification and restatement in terms of modern knowledge to take account of (a) what is now known of the forms in which lipid is transported in the blood; (b) the known association of risk factors with the occurrence of athero-sclerosis; and (c) the mechanisms whereby some of the more important risk factors operate at the level of the arterial wall. The present communication will deal with the role of lipoproteins in the evolution of the atherosclerotic plaque. The accompanying communications will deal with the influence of other factors on this process.

Evidence implicating lipoproteins in atherogenesis

Indirect (correlative) evidence relating the serum levels of low-density (LDL), intermediate (IDL) and very-low density lipoproteins (VLDL) - hereafter collectively referred to as total low-density lipoproteins, or TLDL - with the occurrence of atherosclerosis with respect to age, sex and geographical location in health, and also with respect to diseases accompanied by hyperlipidaemia, has been reviewed elsewhere (3, 5-7).

Direct evidence of the implication of TLDL in atherosclerotic
lesions has been obtained by (i) the use of istopically-labelled
lipoproteins; (ii) immuno-histochemical techniques (iii) examination of arterial extracts and (iv) by a combination of (ii) and
(iii).

(i) Use of radioisotopically labelled lipoprotein. In certain
    instances in which autologous TLDL, radioactively
    labelled in its protein moiety (apolipoprotein) has been
    administered in vivo for turnover studies, and subsequent
    autopsy examination has been possible, it has been shown
    (8, 9) that atherosclerotic vessels accumulate radioactivity
    at a higher level per gram of tissue than other organs,
    suggesting direct entry of the labelled lipoprotein into the
    affected sites.

(ii) Immunohistochemical evidence. Using the technique of
     immunofluorescence and antisera to: intact LDL (subsequently
     shown to react mainly or exclusively with apolipoprotein B,
     an antigenic component common to LDL, I.D.L. and VLDL); and to
     intact high-density lipoprotein (HDL), the presence of TLDL
     but not HDL in atherosclerotic lesions was inferred by several
     authors (for references, see Ref.31. It was observed that the
     distribution of specific fluorescence for apolipoprotein B in
     lesions corresponded exactly with that of extracellular lipid
     (as delineated by conventional fat stains) when the same
     section was examined sequentially by both techniques (10).
     Antibodies to lipoproteins are directed mainly or exclusively
     against the protein part of the molecule, whereas fat "stains"
     such as Oil red O and the Sudan dyes are selectively soluble
     in lipids but unreactive with proteins. Where precise identity
     of distribution between lipid "staining" and immunochemical
     reaction for TLDL was found, therefore, it appeared reasonable
     to infer that the intact lipoprotein molecule must be present.

     On the other hand, in relation to fat-filled cells within
     lesions, it was found that some such cells gave bright specific
     fluorescence for apolipoprotein B, others reacted weakly and
     some failed to react. On the same basis as above, it was
     inferred that full immunoreactivity was compatible with the
     uptake of intact lipoprotein by phagocytic cells whereas
     relative lack, or loss, of immunoreactivity was suggestive of
     progressive degradation of the lipoprotein molecule by
     intracellular proteases. These, it was considered, first
     digest the apolipoprotein (site of antigenicity) leaving a
     lipid residue in the cell (also possibly contributed to by
     local lipid synthesis of triglyceride and phosopholipid)
     detectable by conventional fat "stains". The appearance of

fat-filled cells was thus thought to indicate the activity of
a cellular removal mechanism.

Using antisera specific for other individual plasma
proteins the only one found (11) significantly to be related
to atherosclerotic lesions was fibrinogen, in confirmation
of other observations (12-14).

In later work, antisera to additional antigenic
components of lipoproteins have been employed. These have
included antisera to apolipoprotein C and to the arginine-
rich peptide (components present in IDL and VLDL) and to the
peptide characteristic of Lp(a) lipoprotein, a variant
lipoprotein easily detectable in the serum of some persons
but not others (15-17). Identity of distribution in plaques
was found between that of apolipoproteins B and C or ARP, and
also between apolipoproteins B and Lp(a), suggesting that
lipoproteins even larger in molecular size than LDL itself
participate in atherogenesis (also see below).

This suggestion was supported by immunohistochemical
examination of plaques using antiserum to apolipoprotein B
labelled with horse-radish peroxidase instead of, or in
parallel with, fluorescein-labelled antisera (18, 19).
Antibodies labelled with horse-radish peroxidase and bound to
TLDL in tissues, when treated with diaminobenzidine, give
rise to a brown reaction product allowing the localisation of
the lipoprotein in tissue sections in the light microscope.
Following treatment with osmic acid, an electron-dense reaction
product is obtained allowing localisation also in the electron
microscope. On using this technique on plaques from the aorta,
basilar, coeliac axis and splenic arteries (19) areas
containing pools of amorphous lipid (reactive also with
fluorescein-labelled anti-apolipoprotein B) in the interstitial
matrix were examined. In the electron microscope large numbers
of spherical particles coated by electron dense reaction
product were seen. In some instances, the plane of section
was such as to show granules of reaction product arranged in
circles around a core of greyish lipid material, as though
surrounding the circumference in section of a sphere. The
internal diameters of these structures was found to vary
between about 22 and 48 nm. This corresponds to the range of
molecular diameters found for individual lipoprotein molecules
when isolated and purified fractions of LDL, IDL and VLDL are
examined in the electron microscope. (20) It would seem
therefore, from direct visualisation, that molecules within
the whole size range encompassed by TLDL can participate in
lesions.

(iii) <u>Examination of arterial extracts.</u> Extracts of the arterial
      wall prepared by homogenization, sonication, or by the use
      of a tissue-press have allowed the identification of
      lipoproteins in the tissue fluids by immunoprecipitation
      (20-22). Using a technique of electrophoresing the proteins
      in the interstitial fluid of arterial intima directly into
      antiserum-containing gels, it has been demonstrated (23-25)
      that the amount of TLDL recoverable from the intima is
      related to the level of serum lipids and lipoproteins in
      life. It has also been shown that fibrinogen is similarly
      recoverable from intima and that a proportion of this is
       clottable with thrombin (25).

(iv) <u>Combined techniques.</u>   On applying the method of two-
     dimensional electrophoresis to remove and characterise the
     proteins present in the arterial intima, it could be
     demonstrated that almost all the proteins present in plasma
     are present in the interstitial fluid. But if intima from
     atherosclerotic lesions is thus extracted until no further
     protein emerges and the <u>same</u> intimal sample is then examined
     by conventional histology and by immunofluorescence, then a
     "firmly bound" fraction of lipid and lipoprotein is still
     demonstrable in the tissues (26).

     On applying some of the above-mentioned techniques to areas
unaffected by visible plaque formation, the results obtained vary
with the age of the subject in a manner suggesting variability of
permeability of the arterial wall, with age, to plasma proteins of
differing molecular size (3). Other factors causing selective or
diffuse alteration of vascular permeability under various
pathological circumstances are discussed in Lecture II.

     The evidence discussed so far may be summarised as indicating
that the development of atherosclerotic lesions depends not only
upon the entry of plasma proteins into areas of increased
permeability, but also upon the selective binding at such sites of
these lipoproteins (and fibrinogen) to components of the arterial
wall.

<u>Mechanism, sites of binding, and effect of entrapment of TLDL in
the arterial wall</u>

     The mechanisms variously proposed for retention of the 'firmly
bound' fraction of TLDL in the arterial wall have been previously
reviewed and discussed (3, 27). Evidence for the view gaining
widest acceptance, namely that TLDL and fibrinogen are retained in
the arterial wall by their interaction with the sulphated
glycosaminoglycans (S-GAG's) or proteoglycans (S-PG's) of the
connective tissues of the arterial wall, can be summarised as

follows :

     (i) In relation to lipoproteins, selectivity of retention of
TLDL but not HDL implies that it is the <u>nature of the vehicle</u> for
lipids (some aspect of the physicochemical characteristics of TLDL)
rather than the nature of the lipids being carried, which determines
initial localisation. Since fibrinogen is also present in lesions,
one might suspect both TLDL and fibrinogen to share some common
characteristic of reactivity. In this context it seems of significance that among the plasma proteins, only TLDL and fibrinogen
selectively form insoluble complexes (co-acervates) on the addition
of charged polysaccharides, S-GAG's, or S-PG's derived from the
arterial wall, under <u>in vitro</u> conditions (for references, see
Refs 3 and 27).

     (ii) In the intact artery, histochemical methods and
radiosulphate uptake show a significant correlation between the
distribution of S-GAG's or S-PG's and of LDL and fibrinogen in
plaques (28, 29).

     (iii) Even "mucoid elevations" of the intima - very early
lesions which some authors (3o, 31) hold to precede visible lipid
(lipoprotein) accumulation in the wall - show local accumulation of
S-GAG's. Moreover, many of the stimuli giving rise to increased
vascular permeability also produce a "mesenchymal reaction" (32)
in the subendothelial tissues of arteries, associated with increased
radiosulphate uptake, suggesting increased production of S-GAG's.
If the conditions giving rise to altered permeability also promote
synthesis of the components which are reactive with and precipitate
TLDL and fibrinogen,, it is possible to see how and why selective
entrapment of these proteins is initiated.

     The S-GAG's and S-PG's of connective tissues are intimately
involved in the structural organisation of collagen and elastin.
During fibrillogenesis they provide a framework allowing the
orientation and steric arrangement of developing micro-fibrils to
allow their cross-linking and aggregation. Fine structural studies
(18, 19) using the immunoperoxidase technique have shown that TLDL
molecules can be identified on the outer surfaces of bundles of
micro-fibrils forming the collagen fibre seen in the light
microscope and thus conforming with the "peri-fibrous" distribution
(33) seen by conventional microscopy. But TLDL binding to
individual fibrils of native (mature) collagen and of elastin was
also demonstrable.

     In advanced plaques, TLDL binding was also found in relation
to an 'abnormal' polymorphic form of collagen, known as fibrous
long-spacing collagen (FLSC). In this instance, the TLDL was
bound at the prominent transverse bands occurring at intervals of

about 120 nm along these along these structures. It is precisely
at these sites where interfibrillar bridging by S-GAG's has been
shown to occur (34).

Significance of binding of TLDL

The entrainment of TLDL in the intimal gel often evokes a local
cellular reaction. This may serve, as previously suggested, as a
removal mechanism. But alternatively, the degradation of the
lipoprotein and release from it of free and esterified cholesterol,
phospholipids and other components may stimulate further cellular
reaction and fibrosis. Alternatively a mixture of intact and
degraded lipoprotein, probably contributed to by local triglyceride
and phospholipid synthesis, may result in the accumulation of
"atheromatous gruel" in the plaque, making it liable to ulceration,
superadded thrombosis and the later complications of atherosclerosis.

The finding of TLDL binding to FLSC may also be of significance
since it has been reported (35) that when arterial explants are
maintained in tissue culture, the formation of this 'abnormal' form
of collagen occurs in the presence of hyperlipidaemic but not
normolipidaemic serum in the medium. The tensile strength of FLSC
is likely to be less than that of 'normal' collagen so that its
occurrence may be conducive to arterial dilatation, aneurysm
formation and other complications associated with reduced tensile
strength of the arterial wall (19)

REFERENCES

1. Thoma, R. (1883) Virchows Arch. Pathol. Anat. Physiol.,
        93, 443.

2. Rokitansky, C.von (1952) In A Manual of Pathological
        Anatomy (translated by G.E. Day), Vol 4, p.261.
        Sydenham Society, London.

3. Walton, K.W. (1975) Amer. J. Cardiol., 35, 542.

4. Virchow, R. (1862) In Gesammelte Abhandlungen zur
        wissenschaftlichen Medizin, Max Hirsch, Berlin.

5. Walton, K.W. (1969) In The Biological Basis of Medicine
        (Eds., Bittar, E. and Bittar, N.) Vol.6. pp 193-223,
        Academic Press, New York.

6. Walton, K.W. (1973) In Textbook of Geriatric Medicine and
    Gerontology, (Ed. Brocklehurst, J.C.) pp.77-112,
    Churchill Livingstone, Edinburgh.

7. Walton, K.W. (1974) Adv. Drug Research, 9, 55.

8. Walton, K.W., Scott, P.J., Verrier Jones, J., Fletcher, R.F.
    and Whitehead, T.P. (1963) J. Atheroscler. Res., 3, 396.

9. Scott, P.J. and Hurley, P.J. (1970) Atherosclerosis, 11, 77.

10. Walton, K.W., Williamson, N. and Johnson, A.G., (1970)
    J. Pathol., 101, 205.

11. Walton, K.W. and Williamson, N. (1968) J. Atheroscler., 8, 599.

12. Woolf, N. and Crawford, T. (1960) J. Pathol. Bact., 80, 405.

13. Haust, M.D., Wyllie, J.C. and More R.H. (1964) Amer. J. Pathol,
    44, 255.

14. Wyllie, J.C., More, R.H. and Haust, M.D., (1964) J. Pathol.
    Bact., 88, 335.

15. Walton, K.W. In Atherosclerosis III, (Eds. Schettler, G.
    and Weizel, A), pp. 93-95, Springer Verlag, Berlin.

16. Walton, K.W., Morris, C.J. and Pagnan, A. - in preparation.

17. Walton, K.W., Hitchens, J., Magnani, H.N. and Khan, M.K. (1974)
    Atherosclerosis, 20, 323.

18. Hoff, H.F. and Gaubatz, J.W. (1975) Virchows Arch. Pathol.
    Anat. Histol., 369, 111.

19. Walton, K.W. and Morris, C.J. In Proceedings of First Inter-
    national Atherosclerosis Conference, Austrian Association
    for Morphological and Functional Research in Atheroscler-
    osis, Vienna, 1977, (ed. H. Sinzinger) Karger, Basel, in
    press.

20. Ott, H., Lohss, F. and Gergely, J., (1958) Klin. Wschr., 36,383

21. Tracey, R.E., Merchant, E.B. and Kao, V.C. (1961) Circ.Res.,
    9, 472.

22. Klimov A.N., Denishenko, A.D. and Magracheva, E.Y. (1974)
    Atherosclerosis, 19, 243.

23. Smith, E.B. and Slater, R.S. (1971) Biochem. J., 123, 39P.

24.   Smith, E.B. and Slater, R.S. (1972) Lancet, 1, 463.

25   Smith, E.B., Slater R.S. and Hunter, J.A. (1973) Atheroscler-
      osis, 18, 479

26.   Walton, K.W. and Bradby, G.V.H. In Proceedings of Internat-
      ional Workshop-Conference on Atherosclerosis, London,
      Ontario, 1975, (ed. M.D. Haust), Plenum Press, New York,
      in press.

27.   Walton, K.W. (1974) In Atherosclerosis III. (Eds. Schettler,G.
      and Weizel, A.) pp. 215-217, Springer Verlag, Berlin.

28.   Curran, R.C. and Crane, W.A.J. (1962) J. Pathol. Bact. 84,405.

29.   Gerö,S., Virag, S., Bihari-Varga, M., Szekely. J. and
      Feher, J. (1967) In Progress in Biochemical Pharmacology
      (Eds., Kritchevsky, D., Paoletti, R. and Steinberg, D.)
      Vol. 2, p.290, Karger, Basel.

30.   Haust, M.D. (1971) Human Pathol., 2, 1.

31.   Smith, E.B. and Slater, R.S. (1973). In Atherogenesis :
      Initiating Factors. Ciba Foundation Symposia, New Series,
      No.12, p.39, Excerpta Medica, Amsterdam.

32.   Hauss, W.H., Junge-Hülsing, G. and Hollander, H.J. (1962)
      J. Atheroscler. Res., 2, 50.

33.   Smith, E.B., Evans, P.H. and Downham, M.D. (1967)
      J. Atheroscler. Res., 7, 171.

34.   Doyle, B.B., Hukins, D.W.L., Hulmes, D.J.S., Miller, A. and
      Woodhead-Galloway, J. (1975) J. Molec. Biol., 79-99.

35.   Larrue, J. Darel, D., Demond. J. and Bricaud, H.
      Atherosclerosis - in press.

THE INTERACTION BETWEEN HYPERLIPIDAEMIA AND OTHER RISK FACTORS IN THE PATHOGENESIS OF ATHEROSCLEROSIS (LECTURE II)

K.W. Walton

Department of Experimental Pathology
University of Birmingham
Birmingham 15   England

In my first lecture evidence derived from application of a variety of techniques was reviewed which suggests that atherosclerotic lesions arise in general because certain macromolecular plasma proteins (most notably TLDL and fibrinogen) interact with components of the connective tissues of the arterial wall. In the present lecture, evidence will be considered which points to this being a general mechanism which even affects extravascular connective tissues at certain sites.

If a modified insudative mechanism of atherogenesis is provisionally accepted, then the composition of the plasma will clearly be of importance. But it is often objected that the composition of the blood in respect of TLDL and fibrinogen concentrations is uniform throughout the vascular tree and yet lesions occur discontinuously and only at certain sites of predilection in arteries and not in reins. It is also objected that atherosclerosis may be encountered in individuals showing minimal or no elevation of serum lipid levels (1).

In seeking to answer these objections it is necessary:
(a) to distinguish between the factors determining the <u>localisation</u> and those increasing the severity of lesions;
(b) to appreciate how certain factors mutually re-inforce one another in a fashion which is not simply additive;
(c) to seek understanding of the underlying mechanisms by examining sites other than the arterial plaque itself.

It is suggested that the prime factor determining the localisation of lesions is alteration of permeability of the vessel wall at the sites affected.

General considerations relating to variation of vascular permeability

The relative permeability of endothelium by different plasma proteins, and therefore the distribution of given protein between intravascular and extravascular compartments, is probably largely determined by the molecular size and shape of the protein. For example, in the case of a relatively small molecule like serum albumin, isotopic turnover studies have established that $1\frac{1}{2}$ - 2 times as much albumin is distributed extravacularly as is present intravascularly (2,3). On the other hand, in the case of TLDL, similar methods have shown that, of the total body pool, about two-thirds is intravascular and only one-third (much of which is probably accounted for by the liver pool) is extravascular (4-8). Moreover, labelled TLDL molecules introduced into the circulation have been found to reach full equilibrium very slowly with the extravascular pool (8).

These observations lead to the conclusion that TLDL is normally largely confined to the circulation and that demonstration of its presence (and in particular of VLDL, the largest of the size-range of these molecules) in the arterial wall (a site outside the circulation) can be logically regarded as an indication that endothelial permeability is selectively altered at such a site.

Selective alteration of permeability

In experimental animals naturally prone to the development of atherosclerosis, such as the pig, it has been shown repeatedly that areas, in which lesions develop preferentially, at an earlier stage show selective alteration of permeability in allowing the ingress to the same areas of a variety of marker substances such as dye-labelled proteins, colloidal iron, colloidal lipid and even finely particulate material such as ferritin or horse-radish peroxidase (for references, see Ref. 9 and $1\frac{1}{4}$).

With regard to the mechanism of production of localised alteration of permeability at such sites of predilection, it seems reasonable to assume that the arterial wall, like other living connective tissues, can react to stimulation by the local release of histamine, kinins, and other intrinsic mediators of altered permeability. Whether the 'trauma' of haemodynamic stresses at bends, flow-dividers etc. in the arterial tree can operate in this way is difficult to determine in the human. Experimentally, local trauma to elective sites, applied either internally or externally to the artery (for reviews, see 11,12) causes initial oedema. Later, lipid infiltration has been observed even in normolipidaemic animals but is accelerated if hyperlipoproteinaemia is induced (13).

Localised alterations which are widespread in the arterial tree may be immunologically mediated by allergic injury to arteries.

For example, the induction of immune complex disease in lipid-fed
rabbits increases the extent and severity of lipid-filled arterial
injuries (14). However, it has been noted that in such experiments
lesions can be observed at sites not usually involved in animals
subjected to lipid-feeding alone (15). This is probably because
immune complexes have their own characteristic sites of lodgement
(in small arteries) which are not identical with the common sites
of atherosclerotic involvement (in large arteries). While it can
be envisaged that immune complexes (and especially those incorpor-
ating activated complement and involving blood platelets) may
damage endothelium, in personal experiments (Valente and Walton,
unpublished) it has been found difficult to demonstrate the
simultaneous presence of components of such complexes, and of
lipoprotein infiltration in lesions in this experimental model.
This is probably because the rate of removal of the components of
immune complexes is rapid in relation to the time-scale of
production of unequivocal atherosclerotic lesions. However, it is
possible that a human counterpart to the experimental situation is
seen in the increased speed of development and severity of atherosclerosis seen in the vessels of transplanted organs undergoing
rejection and also showing evidence of immune arteritis.

Widespread but segmental alteration of permeability can be
envisaged as occurring in arterial territories involved by
inflammatory processes. This may account for the observed tendency
for atherosclerosis to be more severe when it supervenes in
arterial segments affected by various forms of granulomatous
arteritis such as temporal arteritis, Takayasu's disease or
syphilitic arteritis (16 - 19). Granulomatous arterial disease
also affects the outflow from the wall because of obstruction of
lymphatic drainage (9).

Diffuse alteration of vascular permeability

In the previous lecture, attention was drawn to the generalised
increase in vascular permeability which accompanies ageing. This is
given its fullest expression in the 8th and 9th decades of life
since the thickened and pigmented aortic intima of individuals in
these age groups, even in areas apparently unaffected by discrete
atherosclerotic lesions, shows diffuse sub-intimal infiltration by
lipid in the form of TLDL (20).

In acute experiments, the intravenous injection of histamine
epinephrine and other vaso-active substances produces a generalised
"oedematous intimal reaction' attended later by the insudation of
TLDL into the intima (21).

Hypoxia was suggested by Hueper (22, 23) as a factor inducing
diffuse alteration of vascular permeability so as to allow the
influx into the arterial wall of both exogenous and endogenous

molecules(including the plasma lipids) from the circulation.
Experimentally, the maintenance of lipid-fed hyperlipoproteinaemic
animals at reduced oxygen tensions (24) or in atmospheres containing significant concentration of carbon monoxide (25) accelerates
and intensifies experimental atherosclerosis.

The human counterpart of this is seen in the heavy cigarette
smoker (26) who may show a consistent level of 10-15 per cent of
carboxyhaemglobin in the blood (i.e., a level found to enhance
atherosclerosis in the experimental animal). It seems likely that
this is one of the underlying mechanisms leading to the recognition
of heavy smoking as a risk factor (27). It has been observed that
diffuse sub-intimal lipid (TLDL) infiltration in areas unaffected
by visible raised lesions, and resembling that otherwise seen in
the last decades, is observed precociously in the arteries of young
males who were heavy smokers in life (20).

At the ultrastructural level, in animals subjected to hypoxia,
the injection of vaso-active compounds etc.,damage is demonstrable
to endothelial cells and the opening of endothelial cellular
junctions is seen (28, 29).   In the preceding lecture attention
was drawn to evidence that stimuli of this kind also evoke a
"mesenchymal reaction" in the subendothelial tissues associated
with increased production of sulphated glycosaminoglycans and
proteoglycans which may serve to entrain TLDL and fibrinogen
entering the wall.

Interaction of hyperlipidaemia and altered vascular permeability

Once increased permeability has been established, it can be
envisaged that there would be an increased volume of plasma
permeating in unit time through a given affected segment of the
wall, as well as ingress of the large molecules normally largely
excluded. This would permit TLDL and fibrinogen to enter in higher
total concentrations at the affected site even in the absence of
raised concentrations of these proteins in the circulating plasma,
perhaps explaining why atherosclerosis may be initiated and progress
in spite of normal or only slightly elevated serum lipid levels in
man. However, clearly the process envisaged would be further
accelerated with the concomitant occurrence of hyperlipidaemia
(hyperlipoproteinaemia).

The interaction between increased vascular permeability and
hyperlipidaemia is well illustrated in relation to the pathogenesis
of xanthomata (30).   These are most commonly seen in man in
association with both primary (familial) hyperlipidaemias and the
secondary hyperlipidaemias accompanying diabetes mellitus, the
nephrotic syndrome, hypothyroidism and obstructive biliary disease.
Xanthomata are characteristically distributed at sites subjected to
minor trausma (elbows, knees, buttocks); in skin subjected to

constant creasing or folding (palmar creases, eyelids); in tendons
subjected to friction in the neighbourhood of joints (finger tendons)
or in tendons subjected to constant tensional stress (tendo Achillis)
If treatment is effective in reducing hyperlipidaemia, xanthomata have
been repeatedly noted to decrease in size or to regress, with loss
of the lipid contained in them (for references, see Ref.30).   On
the other hand, in relation to the active progression of lesions in
untreated cases of hyperlipidaemia, it was shown by Scott and
Winterbourn (31) that when autologous radioactively labelled TLDL
was injected into hyperlipidaemic subjects with actively growing
xanthomata symmetrically distributed upon both elbows, greater
localisation of radioactivity occurred in a lesion subjected to
slight but repeated trauma than in a protected lesion on an
immobilised limb.   These results suggested that xamthomata might
arise by leakage of TLDL-rich plasma into the connective tissues,
because of localised increase of vascular permeability due to
the release (occasioned by repeated slight trauma at the affected
sites) of histamine and other vaso-active compounds, with uptake
of the lipid (lipoprotein) by tissue histiocytes.

    This hypothesis was also supported by experiments in rabbits
(30).   In this species xanthomata with histological characteristics closely similar to those of human lesions occur spontaneously in animals maintained on lipid-supplemented diets and with
hyperlipoproteinaemia.   But it was shown that similar lesions
could be induced at elective sites by the repeated injection of
histamine or bradykinin .   No similar lesions could be induced in
normolipidaemic animals.   Using radioactively labelled rabbit
TLDL, or the technique of immunofluorescence, it was shown that
the material in the lesions was indeed rabbit TLDL.   Moreover
it was noted that early induced lesions in animals with moderate
hyperlipoproteinaemia regressed when the injections giving rise
to altered vascular permeability were interrupted.   On the other
hand, lesions in animals with gross hyperlipoproteinaemia were
more rapidly and easily induced, persisted and became indurated.

Hyperlipidaemia and other risk factors

    Epidemiological studies have documented the association of
hyperlipidaemia with the various forms of cardiovascular,
cerebrovascular and peripheral vascular morbidity and mortality
arising from atherosclerosis in the arteries supplying these
various territories.   Since HDL show relatively little variation
in health or disease (9) hyperlipidaemia, in practice, is given
expression in elevation of TLDL manifested as increases in one
or more of its subclasses.

Moderate hyperlipidaemia.   The effects of age, sex, geographic
location, cultural conventions etc upon serum lipid levels and

the incidence of atherosclerosis on a population basis have been
discussed previously (9, 32). Most attention in studies of this
kind has been concerned with the quality and proportion of fat in
the diet in influencing serum levels. More recently, attention
has also turned to the quality of dietary protein (whether from
vegetable or animal sources) and the effect of this upon serum
lipids and lipoproteins (33, 34).

There is evidence that the apolipoproteins of HDL and of TLDL
are synthesized, and the intact lipoproteins assembled, in both the
liver and in the epithelial cells of intestinal mucosa (35, 36).
Nothing is known of the relative contributions made by each site
of synthesis to the total body pools of these lipoproteins, nor is
it known how control of the rate of synthesis at these sites is
exerted, though clearly one would expect dietary variation to exert
effects upon both sites.

Gross hyperlipidaemia.   The accelerated rate of development of
atherosclerosis, the precocious occurrence of a corneal arcus, and
the frequent presence of xanthomata in association with both severe
secondary and with primary (familial) hyperlipidaemias is well
documented (for references see Ref.37).

Table I

Risk of cardiovascular disease (CVD) and patterns of alterations of
lipoproteins in familial hyperlipoproteinaemias. *

Type ≠		Risk of CVD ✝
I	Gross chylomicronaemia; VLDL, IDL and LDL normal or low; deficient lipoprotein lipases	Normal
IIA	Marked increase of LDL	Very high
IIB	Increased LDL with moderate increase of VLDL	High
III	Increase of LDL and IDL ("floating betalipoprotein")	High
IV	Marked increase of VLDL ± increase of LDL	Increased
V	As for IV, with chylomicronaemia	? Increased

* Slightly modified from Walton (9) with permission of Editor and
  publishers of Amer. J. Cardiol.
≠ Based on classification of Beaumont et al (38)
✝ From data of Fredrickson et al (39) and of Slack (40)

In both primary and secondary hyperlipidaemias the high serum
TLDL levels reflect a greatly increased total body pool of these
lipoproteins with an increase in both intravascular and extra-

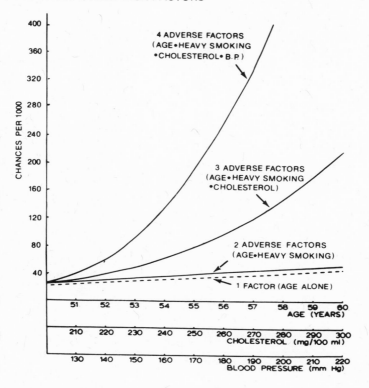

FIGURE. Interaction of risk factors in increasing probability of development of coronary heart disease (C.H.D.) during next 5 years in men aged 50-60 as calculated using multiple logistic equation of Keys et al. (42)

vascular distribution (9). It seems reasonable to infer that the increased extravascular distribution is given expression as TLDL deposits both in the walls of arteries and in the tissues.

    Primary hyperlipidaemias classified according to the system proposed by the World Health Organisation (38) show varying risks of development of cardiovascular disease (see Table I).

    Consideration of this data allows inferences to be drawn about the relative " atherogenic potential" of lipoproteins of differing molecular size. For example, the rare Type I abnormality due to a gross chylomicronaemia associated with deficiency of adipose tissue or hepatic lipoprotein lipase is not characterised by an accelerated development of atherosclerosis. The

Table II

Possible Mechanisms Whereby "Risk Factors" Exert Their
Effect at the Level of the Arterial Wall

Risk Factors	Possible Mechanism
1. Age,smoking (hypoxia), extrinsic or intrinsic vasoactive agents	Increased vascular permeability (increased inflow)+"non-specific mesenchymal response"
2. Age,sex,high fat intake as proportion of total calories	Moderate hyperlipoproteinemia (increased input of total low density lipoprotein even with unchanged volume of inflow.
3. Primary or secondary hyperlipoproteinemia	Gross hyperlipoproteinemia (effect as in 2 but more prounced)
4. Hypertension	Further increase of inflow (effect especially marked in hyperlipidemia)
5. External injury to vessel or coexistent arteritis	Inflammation of arterial wall increasing vascular permeability; obliteration of lymphatic vessels reducing outflow

functional defect in such individuals is an inability to degrade
chylomicra into smaller lipoproteins. One might therefore infer
that intact chylomicra are too large to enter the arterial wall.
At the other end of the scale, it might be expected that the
smallest lipoprotein molecules of the TLDL family (the LDL class)
might penetrate most easily through the endothelial barrier.
This accords with the data of Slack (40) who found that men with
the Type IIA or IIB abnormality (preponderant increase of LDL)
had an earlier onset of ischaemic heart disease than did men with
Types III, IV or V abnormalities. Nevertheless, the incidence
rate was high even in persons with a Type IV or V pattern
(predominant increase of VLDL). Visual evidence that molecules
larger than LDL do, in fact, occur in atherosclerotic lesions was
discussed in Lecture I.

Hypertension

Increased blood pressure is generally acknowledged to
accelerate arterial disease in man (27) and in experimental animals.
The manner in which pressure effects are especially evident at
certain sites (for example in the pulmonary artery in pulmonary
hypertension, in the proximal segment in coarctation of the aorta,
in the mitral and aortic valves and in autologous vein grafts)

has been previously discussed (9) and will be dealt with in more detail in Lecture III.

Interaction of hyperlipidaemia and other factors

The manner in which risk factors reinforce one another in a geometric rather than an arithmetic summation of effect has been given mathematical expression. Calculations based on one of the formulae (42) put forward are expressed graphically in the accompanying figure. The manner in which some of the factors probably exert their effect on the arterial wall (summarised in Table II) and summate in so doing, has already been alluded to in this Lecture. Further human and experimental models illustrating the interaction will be discussed in Lecture III.

REFERENCES

1. Pickering, G.W. (1969) Brit.Med.J., 1, 517.
2. Cohen,S.,Freeman,T.and MacFarlane,A.S.(1961) Clin.Sci.,20, 161.
3. Takeda,Y. and Reeve,E.B.(1963) J.Lab.Clin.Med.,61,183.
4. Scott,P.J.and Hurley,P.J.(1969)Israel J.Med.Sci,5,631.
5. Scott,P.J.and Hurley,P.J.(1970) Atherosclerosis,11,77.
6. Walton,K.W.,Scott,P.J.,Dykes,P.W.and Davies,J.W.L.(1965) Clin.Sci.,29,217.
7. Scott,P.J.,Dykes,P.W.,Davis,J.and Walton,K.W.(1963) In Biochemical Problems of Lipids (ed.,Frazer,A.C.) pp.318-324, Amsterdam,Elsevier.
8. Hurley,P.J. and Scott, P.J. (1970) Atherosclerosis, 11, 51.
9. Walton,K.W.(1975) Amer.J.Cardiol., 35, 542.
10. Nagy,Z.(1974) In Arterial Lesions and Arteriosclerosis (ed.Jellinek,H.) pp.110-137, London, Plenum Press.
11. Adams, C.W.M.(1967) Vascular Histochemistry in Relation to the Chemical and Structural Pathology of Cardiovascular Disease, p.74,London, Lloyd-Luke .
12. Constantinides,P.(1965) Experimental Atherosclerosis, pp.27-51, Amsterdam, Elsevier.
13. Björkerud,S. and Bondjers,G.(1973) Nutr.Metab. 15, 27.
14. Minick, C.R., Murphy, G.E. and Campbell,W.G.(1966) J.Exp.Med., 124, 635.
15. Minick, C.R.,(1976) Ann.N.Y.Acad.Sci., 275, 210.
16. Ask-Upmark,E. and Fajers,C-M.,(1956) Acta Med.Scand. 155, 275.
17. Shrire,V. and Anderson,R.A.(1964) Quart.J.Med.,33,439.
18. Turnbull, H.M.(1915) Quart.J.Med., 8, 201.
19. Nichols, E.F. (1940) Am.Intern.Med., 14, 960.
20. Walton,K.W. (1974) Adv.Drug Research, 9, 55.
21. Shimamoto, T. (1974) In Atherosclerosis III. (eds.Schetter,G. and Weizel,A.) pp.68-82, Berlin, Springer Verlag.
22. Hueper,W.C.(1944) Arch.Pathol.38, 162,245,350; (1945) Arch. Pathol., 39, 57, 117, 187.

23. Hueper W.C.(1956) Amer.J.Clin.Pathol., 26, 559
24. Myaskinov, A.L. (1958) Circulation, 17, 99.
25. Astrup, P., Kjeldsen,K and Wastrup,J.(1967) J.Atheroscler. Res.,
    7, 343.
26. Kjeldsen,K. In Atherosclerosis:Proceedings of the Second Inter-
    national Symposium (ed.Jones R.J.) pp.378-381, Berlin,
    Springer Verlag.
27. Dawber,T.R.,Kannel,W.B.,Revotskie, N. and Kagan, A.(1962)
    Proc.R.Soc.Med.,55, 265.
28. Constantinides, P. and Robinson,M.(1969), 88, 99, 106, 113.
29. Kerenyi, T. and Hüttner, I. (1974) In Arterial Lesions and
    Arteriosclerosis (ed.Jellinek,H.) pp.91-102, London,
    Plenum Press.
30. Walton,K.W., Thomas C. and Dunkerley,D.J.(1973) J.Pathol.,109,
    271.
31. Scott, P.J. and Winterbourn, C.C. (1967) J. Atheroscler.Res.
    7, 207.
32. Walton, K.W. (1969) In The Biological Basis of Medicine,
    (eds.Bittar,E. and Bittar,N.) Vol.6, pp.193-223, New York,
    Academic Press.
33. Hamilton,R.M.G. and Carroll,K.K.(1972) Atherosclerosis, 24,47.
34. Sirtori, C.R., Agradi, E., Conti,F., Mantero, O. and Gatti, E.
    (1977). Lancet, 1, 275.
35. Hamilton,R.H. (1972) Adv.Exper.Med.Biol., 26, 7.
36. Walton, K.W., Morris, C.J. and Bradby, G.V.H. (1977) Nature
    (Lond.) in press.
37. Walton, K.W. In Textbook of Geriatric Medicine and Gerontology,
    (ed., Brocklehurst,J.C.) pp.77-112, Edinburgh, Churchill
    Livingstone.
38. Beaumont, J.L., Carlson, L.A., Cooper, G.R., Fejfar, Z.,
    Fredrickson, D.S. and Strasser, T. (1970) Bull.Wld.Hlth.Org.,
    43, 891.
39. Fredrickson, D.S., Levy, R.I. and Lees, R.S. (1967) New Eng.
    J.Med., 276.
40. Slack, J. (1969) Lancet, 2, 138¼.
41. Hüttner, I. (1974) In Arterial Lesions and Arteriosclerosis
    (ed.Jellinek, H.) pp.128-134, London, Plenum Press.
42. Keys,A., Aravanis, C., Blackburn, H., Buchem, F.S.P.van,
    Buzina,R., Djordjevic, B.S., Fidanza, F., Karvonen, M.J.,
    Menotti, A., Puddu, V. and Taylor, H.L. Circulation, 1972,
    45, 815

MODELS FOR THE STUDY OF TISSUE DEPOSITION AND SYNTHESIS

OF LIPOPROTEINS (LECTURE III)

K.W. Walton

Department of Experimental Pathology

University of Birmingham,   England

Experimental animal models

Efforts have been made, using a wide variety of animal species, to produce an acceptable animal model of human atherosclerosis. No single model is accepted universally as reproducing all aspects of the condition in humans (for review, see Ref.1) although primate models, and especially those involving the apes, such as the chimpanzees, probably approach this ideal most closely because both the lipid composition (2) and the immunological characteristics (3) of their lipoproteins are very similar to those of man.

However, because of the expense involved in obtaining and maintaining primates and also because of restrictions imposed by the Governments of some countries on the use of certain genera of primates, many laboratories have perforce to work with cheaper and more convenient laboratory animals. It is now becoming recognized that certain species in which the principal lipid-carrying protein is a high-density lipoprotein (dogs, rats, hamsters) are resistant to the induction of experimental atherosclerosis. On the other hand, in species in which the principal lipid-carrying proteins are low-density lipoproteins (as in man) dietary manipulation can be shown to increase the level of these lipoproteins in the blood and to be attended by the development of arterial lesions. Among this latter group is the rabbit. From the limited view-point of examining, under experimental conditions, whether an insudative mechanism operates in atherogenesis in species other than man, it is worthwhile to re-evaluate the validity of use of the lipid-fed rabbit as an experimental model.

Experimental atherosclerosis in the rabbit

It has been shown that whether rabbits are maintained on
cholesterol-supplemented diets, or on semi-synthetic diets
containing fat (e.g. beef lard) without added cholesterol, the
animals develop not only a hypercholesterolaemia, but also a
hyperlipoproteinaemia characterised by an increase mainly of TLDL.
The distribution of density sub-classes within the TLDL varies with
the nature of the diet.   In the cholesterol-fed animal there is
a marked increase of a lipoprotein with VLDL-like density-
characteristics but which differs from normal rabbit VLDL in
containing a much higher proportion of cholesterol than normal (4).
The protein moiety of this 'abnormal' lipoprotein also differs from
normal rabbit VLDL in containing an unusually high proportion of
a peptide probably analgous with human arginine-rich peptide
(apolipoprotein E) according to some authors (5 - 7) and in its
large and relatively uniform size as visualized by electron
microscopy (7).   Cholesterol-fed animals often show a reciprocal
reduction in high-density lipoproteins (4).   On the other hand,
beef fat fed rabbits show a more moderate hyperlipidaemia, attended
by an increase of both HDL and TLDL but the latter show a distribution of density-classes within the TLDL which more closely
approximates to that in normal rabbits (4).

Despite these differences, it was established in animals on
both the above dietary regimes that the arterial lesions, when
examined by the technique of immunofluorescence, showed the presence
of rabbit TLDL (8) distributed in a manner closely similar to that
previously found in human lesions using a similar technique (9).
Moreover, when isotopically-labelled rabbit TLDL was injected
intravenously, the labelled molecule was found to localise in the
aorta and other lesions in these animals (4).   This model was
found also to be suitable for study of the mechanism involved in
the deposition of lipid (lipoprotein) into certain intravascular
sites to give rise to lesions which have long been known to be
broadly correlated with the occurrence of human atherosclerosis,
namely xanthomatosis and corneal arcus formation (4.10,11).

The Dietary factor in atherosclerosis

In the rabbit model, hyperlipidaemia is induced solely by
dietary manipulation.   At the height of the hyperlipidaemia so
induced there is evidence that the synthesis of cholesterol and
phospholipids by the liver is greatly reduced (for references see
Ref.12) and that synthesis of lipoproteins by parenchymal liver
cells is similarly reduced (8).   It may be inferred therefore that
the great bulk of the circulating lipoproteins is of intestinal
origin.

In man, epidemiological observations suggest that the mean

level of circulating lipids and lipoproteins is influenced by the
proportion of the dietary intake which is in the form of fat (13)
and, in particular, of saturated fat.   But in normal man, unlike
the rabbit, there appears to be a control mechanism, safeguarding
against grossly excessive fat absorption from the gut since,
following a very large fat-meal, only a proportion of the fat is
absorbed, the remainder being excreted as faecal fat.

    While there is good evidence for the transformation of
absorbed fat into lipoprotein by intestinal epithelial cells in
the rat (14) there has hitherto been little direct evidence that
intestinal lipoprotein synthesis also occurs in man.   Recent work
by the author and his collaborators (15) using immunofluorescence
established that apolipoproteins A, B, C and E were all demonstrable
both in liver cells and in intestinal epithelial cells from normal
subjects.   Ultrastructrual studies, using the immunoperoxidase
technique and a labelled antiserum to apolipoprotein B suggested
that in intestinal epithelial cells the smooth endoplasmic
reticulum, the Golgi apparatus and secretory vesicles derived from
the latter, served as a pathway for the assembly, transport and
secretion of the serum lipoproteins into intestinal lymph.   An
essentially similar intracellular pathway has been described for
liver cells (16).

    Currently virtually nothing is known of the relative
contributions made by liver and intestinal synthesis to the total
body pool and the circulating level of lipoproteins in man.   But
it is possible that in certain human pathological conditions
(perhaps particularly primary hypertriglyceridaemias characterised
by an increase of endogenous VLDL) the contribution by the
intestinal site may be excessive.   Hyperlipidaemias of this kind
are relatively common and associated with an increased incidence
of atherosclerosis (17).

Contributory factors accelerating insudation of lipoproteins

    Increased blood pressure has been clearly established as a
factor accelerating arterial disease in man (18) and in experimental
animals (19).   It can be envisaged that, once selective alteration
of permeability (a localised "leak") has been established in an
artery, the volume of plasma transuding through the affected
segment would be increased in unit time.   In this way the total
concentration of potentially reactive molecules (TLDL and fibrinogen)
entering the segment in unit time would be increased even if the
plasma levels of these components were normal.   The opportunity
for their entrapment in the gel would be thus increased.   But
clearly the effect would be greatly magnified by concomitant
hyperlipidaemia.   This mechanism can be evoked to explain the
clinical observation that hypertension increases the risk of

development of cardiovascular disease but that the risk is magnified in subjects with elevated levels of blood lipids.

The effect is particularly marked at certain sites.   For
example, atherosclerosis of the pulmonary artery in man is only
seen in association with pulmonary hypertension (20) and in
co-arctation of the aorta is sharply delimited to the proximal
(high-pressure) segment.   Attention has previously been drawn
(21) to the significance of atherosclerosis of the heart valves.
This is seen only in the valves in the left side of the heart
(mitral and aortic valves) which are at arterial pressure and
not in the valves of the right side of the heart which are at
venous pressure.   Moreover the distribution of lipid (lipoprotein)
infiltration is of significance.   In the mitral valve, lipoprotein
infiltration is confined to the ventricular half of the valve (i.e.,
to the surface exposed to high pressure during closure of the valve
with systole).   In the aortic valve, lipoprotein infiltration is
maximal at the base of the pocket formed by each cusp (i.e. to the
point of maximal pressure from backflow during diastole).

While atherosclerosis is normally found only in arteries (high
pressure vessels) and not in veins (low pressure vessels), when
autologous vein grafts are used to bypass occluded vessels and
become exposed to arterial pressure, lipoprotein infiltration,
closely resembling that seen in atherosclerosis of arteries, can be
shown to occur (22).   Contrariwise, arteries normally exposed to
systemic pressure do not develop atherosclerosis if, owing to
developmental abnormalities, they are exposed only to lower
pressures.   This is well seen in the case of the rare developmental
abnormality affecting the left descending coronary artery in which
this vessel arises from the pulmonary trunk instead of the aorta (23).

Summary of current views of atherogenesis

It is submitted that the evidence discussed in Lecture I is
overwhelmingly in favour of the lipid present in atherosclerotic
lesions arising from 'insudation' of TLDL into the arterial wall
rather than from 'fatty degeneration' of intrinsic components of
the wall, or from breakdown of the components of thrombi deposited
on the surface and incorporated into the wall.

In Lecture II, it was proposed that increase of the relative
permeability of the arterial wall by TLDL is probably the source
of origin of atherosclerotic lesions and probably also of certain
extravascular lesions broadly associated with atherosclerosis.
Localised   alteration of permeability is probably the determinant
of the topographic distribution of lesions but diffuse increase
in vascular permeability increases the severity of lesions.   Since
the arterial wall is living connective tissue, it reacts at sites

of altered permeability by increased mesenchymal activity. This gives rise to increased synthesis of components of the ground substance which entrap and give rise to firm binding of TLDL and fibrinogen. ⑦ Hyperlipidaemia increases the rate and severity of this process but is by no means the only determinant since hypertension, changes in the arterial wall caused by trauma, arterial disease etc., also contribute. The available evidence suggests that where several factors operate, they act synergistically to produce a summation which is geometric rather than simply arithmetic.

In Lecture III it was shown that animal models, including the classic but disputed model of the lipid-fed rabbit, can be used to validate some of the views expressed above. Evidence was also presented for lipoprotein synthesis in man both by the liver and by the intestine and attention was drawn to the need to define the contribution made by each site of synthesis to the total body pool of TLDL and thus to hyperlipidaemia and the increased risk of atherosclerosis. It was also shown that certain aspects of atherogenesis can be illuminated by examining sites other than the arterial plaque itself, in man, and this was illustrated by a number of examples relating to the effect of increased blood pressure.

REFERENCES

1. Wissler, R.W. and Vesselinovitch, D. (1974) In Atherosclerosis III (eds. Schettler, G. and Weizel, A.) pp. 319-325, Springer Verlag, Berlin.

2. Blaton, V. and Peeters, H. (1976) Adv. Exper. Med. Biol., 67, 33.

3. Walton, K.W., Hitchens, J., Valente, A.J. and Blaton, V. (1977) Nature (Lond.) - in press.

4. Walton K.W., Thomas C. and Dunkerley, D.J. (1973) J. Pathol., 109, 271.

5. Shore, B. and Shore V. (1960) J. Lipid Res., 1, 321.

6. Camejo, G., Bosch, V., Arreaza, C. and Mendez, H.C., (1973) J. Lipid Res., 14, 61.

7. Camejo, G., Bosch, V. and Lopez, A. (1974) Atherosclerosis, 19, 139.

8. Walton, K.W., Dunkerley, D.J., Johnson, A.G., Khan, M.K., Morris, C.J. and Watts, R.B. (1976) Atherosclerosis, 23. 117.

9. Walton, K.W. and Williamson, N. (1968)J. Atheroscler.Res.8,599.

10. Walton, K.W. (1973) Nutr. Metab., 15, 37.

11. Walton, K.W. and Dunkerley, D.J. (1974) J. Pathol., 114, 217.

12. Shore B. and Shore V. (1976) Adv. Exper.Med. Biol., 67, 123.

13. Keys, A. (1963) In Atherosclerosis and its Origin (Ed. Sandler, M. and Bourne, G.H.) pp.263-299. Academic Press, New York.

14. Windmueller, H.A. and Spaeth, A.E. (1972) J. Lipid Res.,13, 92.

15. Walton, K.W., Morris, C.J., Bradby, G.V.H. and Pagnan, A. (1977) Nature (Lond) - in press

16. Hamilton, R.H. (1972) Adv. Exper. Med. Biol., 26, 7.

17. Carlson, L.A. and Böttiger, L.E. (1972) Lancet, 1, 865

18. Dawber, T.R., Kannel, W.B., Revotskie, N. and Kagan, A. (1962) Proc. R. Soc. Med., 55, 265.

19. Heptinstall, R.H., Barkley, H. and Porter, K.A. (1958) Angiology, 9, 84.

20. Heath, D., Wood, E.H., Dushane, J.W. and Edwards, J.E. (1960) Lab. Invest., 9, 259.

21. Walton, K.W., Williamson, N. and Johnson, A.G. (1970) J. Pathol., 101, 205.

22. Walton, K.W., Slaney, O.F. and Ashton, F. - in preparation.

23. Burch, G.E. and De Pasquale, N.P. (1964) Amer. J. Med., 37, 159.

DIETARY EFFECTS ON LIPOPROTEIN COMPOSITION

Cesare R. Sirtori, Gian Carlo Ghiselli and
M. Rosa Lovati

Center E. Grossi Paoletti for the Study of
Metabolic Disease and Hyperlipidemias, University
of Milano - 20129 Milano, Italy

INTRODUCTION

Qualitative and quantitative dietary changes can markedly affect plasma lipid levels. Variations of lipidemia have an obvious counterpart in changes of lipoprotein levels and composition. In spite of the large number of studies evaluating dietary effects on plasma lipid levels, clinical studies on the relative composition and structure of lipoproteins following dietary changes are relatively scarce. Numerous studies report, on the other hand, changes of lipoprotein composition and structure in animals fed different diets; this is particularly true for diets used with the purpose of inducing experimental atherosclerosis.

This report will analyze available information on lipoprotein composition following dietary changes, expecially aimed to modify plasma triglyceride and cholesterol levels. In both cases, a separate consideration is given to diets which induce increases of plasma triglyceride and/or cholesterol, and to diets which reverse these situations in clinical syndromes.

LIPOPROTEIN CHANGES IN DIET INDUCED HYPERTRIGLYCERIDEMIAS

Increases of triglycerides (TG) and particularly of TG rich lipoprotein fractions, i.e. chylomicrons and very low density lipoproteins (VLDL) are induced by diets

exceedingly rich in calories, carbohydrates (1), ethanol
(2) and by combinations of these. Addition of fat to
ethanol or to carbohydrates worsens diet induced hypertriglyceridemias.

Carbohydrates

Hypertriglyceridemias, particularly type IV hyperlipoproteinemia, may be secondary to diets containing
an excessive amount of carbohydrates (CHO)(1). This
situation may be reproduced in animals and in man. The
resulting clinical syndrome is similar to that found in
patients with hyperlipoproteinemias.

After induction with CHO rich diets, the resulting
excess of plasma TG is circulated as part of the VLDL
molecules,assembled and secreted primarily by the liver.
Studies on the mechanism of the changes in plasma TG
transport following CHO induction, have demonstrated a
significantly altered VLDL (d<1.006) composition in normal
volunteers and patients with hyperlipidemia (3). Switching
from a diet containing 45% of CHO with 15% of protein, to
a diet containing 80% CHO and 20% protein, results in all
cases in increased plasma TG, on the average about doubled,
without significant changes of plasma cholesterol levels.
TG content of VLDL (Table 1) is about triplicated and
the ratio of TG to protein more than doubled. These
results are consistent with the hypothesis (4) that the
liver may change the pattern of secretion of TG rich particles following CHO induction. They also indicate that
VLDL composition may significantly differ relative to the
dietary habits and particularly to the CHO intake.

The apoprotein composition of VLDL, according to
Carlson and Ballantyne (5), may be altered in the presence
of an increased TG/protein ratio in VLDL particles. In
particular, significant changes may be observed in the
relative proportion of apoproteins CII and CIII. In
case of markedly hypertriglyceridemic VLDL (i.e.,in
type V hyperlipoproteinemia) an almost total absence of
CII is found, whereas CII normally prevails over CIII.
These observations confirm the hypothetical role of CII
in activating lipoprotein lipase (6).

After CHO induction in normal man, Schonfeld et al.
(7) recently described, however, increased CII and decreased CIII 2 in both chylomicrons and VLDL. Apo Al
was markedly decreased in HDL, which also become en-

Table I

CHANGES IN VLDL COMPOSITION FOLLOWING CARBOHYDRATE INDUC-
TION IN MAN (3).

Diet:	CHO 45%	CHO 80%
	Prot. 15%	*Prot.* 20%

PLASMA

$\overline{X}$ TG	214.7 mg/dl	534.2
$\overline{X}$ Chol.	235.2	235.3
preβ	34.6%	61.7

VLDL (d < 1.006)

$\overline{X}$ TG	137.4 mg/dl	438.0
$\overline{X}$ Chol.	31.1	74.2
Prot.	44.4	62.1
TG/P	3.09	7.05
C/P	0.70	1.19
TG/C	4.40 (3-6.7)	5.90 (4.4-7.5)

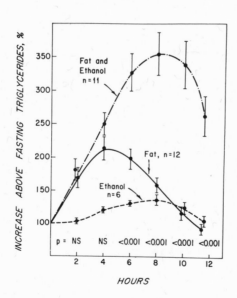

Figure 1: Plasma triglyceride changes following ethanol, fat (100 g of corn oil) and fat + ethanol in healthy volunteers (2).

riched with triglycerides. In normals, therefore, apoprotein changes are indicative of an improved metabolic
clearance; it is possible, on the other hand, that in
patients with hypertriglyceridemia, CHO induction may
cause secretion of VLDL particles with reduced activator
apoproteins.

Lipoprotein changes following CHO induction, in
conclusion, are characterized by TG enrichment of VLDL
and possibly other lipoprotein fractions, with apoprotein
changes normally consistent with an improved metabolic
clearance of the newly secreted lipoproteins. Data on
apoprotein composition in patients with hypertriglyceridemia, suggests that the situation may be different in
these. Thus, the test of CHO induction may be a useful
tool for the diagnosis of the dietary origin of hypertriglyceridemias.

Ethanol

Ethanol represents a very potent inducer of hypertriglyceridemia. This effect of ethanol may be particularly striking in selected patients (8), predisposed
to this type of hypertriglyceridemia, but may be found
in practically all individuals following a load of ethanol
(9).

Several studies have dealt with lipoprotein composition in ethanol-induced hypertriglyceridemia. Wilson
et al. (2) showed that ethanol administration to volunteers induces a 50% rise of plasma TG, and that additional fat (100 g of corn oil) causes a dramatic increase
of triglyceridemia (Fig. 1). No significant changes of
postheparin lipolytic activity (PHLA) and of plasma glu-
cose and insulin are observed, whereas both dietary
treatments, alone and in combination, significantly increase plasma FFA.

Analysis of plasma lipoproteins in patients following
fat and ethanol, demonstrates significant increases of
both chylomicrons (Sf > 400) and VLDL (Sf 20-400). Subfractionation of lipoproteins by analytical ultracentrifuge shows that (Tab. 2) lipoproteins of Sf 100-400
and 20-100 are significantly increased by ethanol and fat,
whereas lipoproteins Sf 12-20 (corresponding to IDL) and
Sf 0-12 (corresponding to LDL) are not modified. An interesting observation is the significant increase of HDL
following this dietary regimen.

The tentative conclusive of this study was that
lipoproteins increased following fat and ethanol are

Table II

LIPOPROTEIN CONCENTRATIONS AFTER INGESTION OF FAT AND ETHANOL (2)[+]

Time after ingestion	Lipoprotein concentration (mg/100 ml)				
	$S_{f}100-400$	$S_{f}20-100$	$S_{f}12-20$	$S_{f}0-12$	HDL
CONTROL	21.2	46.6	18.6	418	111
4 HOURS	61.5	63.9	15.0	450	149
10 HOURS	58.1	87.2	17.8	372	167

[+] Analytical ultracentrifugation of plasma at density 1.20 g/ml. Lipoprotein classes are expressed in S_{f} 1.063 units.

likely to be all of liver origin, independent of changes
in PHLA (10). This conclusion was, however, not supported
by the fatty acid composition of VLDL after fat and
ethanol, similar to that occurring after fat alone, and
by the ineffectiveness of medium chain triglycerides in
modifying ethanol induced hypertriglyceridemia, also
suggesting a possible impairment of clearing mechanisms.

Somewhat different lipoprotein changes were described by Avogaro and Cazzolato (11) in alcoholic subjects
given a large load of ethanol (180 g in 6 hrs.) and studied
for the following 24 hrs. The acute hypertriglyceridemia
(from 26% to 367% above basal values) was secondary to TG
increases in all lipoprotein fractions, maximal in VLDL,
followed by HDL and LDL. Analyses of lipoprotein distribution by polyacrylamide gel electrophoresis and analytical ultracentrifuge were suggestive of the formation of
an extra-lipoprotein fraction, presumably of d 1.006-1.019
(IDL). Detection of this lipoprotein fraction was variable, although rather constant, among subjects. It was
often found to be associated with the appearance of a
"pre-β" band in the electrophoresis of LDL-HDL ("sinking
pre-β") separated by preparative ultracentrifuge. This
last data is somewhat in contrast to that of Wilson et al.
(2) in patients given ethanol and fat. It is possible
that when ethanol is given alone in toxic doses it may
increase lipoproteins of intermediate density, whereas
when given together with fat, only physiological lipoproteins of very low density are increased. Further apoprotein studies are needed to detect specific apoprotein
components following ethanol.

In conclusion, ethanol administration increases VLDL
and chylomicrons and determines a general TG enrichment
in all lipoproteins. There is disagreement between data
reporting no specific changes of lipoprotein distribution,
and other describing increases of IDL. An increase of
HDL following ethanol, has been reported by both Wilson
et al. (2) and by Avogaro (12). This finding may cast
some doubts on the role of ethanol in atherogenesis.
Whereas, in fact, ethanol is generally considered an
atherogenic agent (13), increases of HDL may provide a
protective effect (14).

HYPOTRIGLYCERIDEMIC DIETS AND LIPOPROTEIN COMPOSITION
AND STRUCTURE

The most common dietary treatment for hypertriglyceridemias is CHO reduction. CHO restricted diets,

similar to those commonly used in therapy of diabetes
mellitus (15), are generally effective on plasma TG levels,
but have been shown to elicit some yet unexplained lipoprotein modifications. In particular, as also observed
after treatments with hypotriglyceridemic drugs (e.g.,
clofibrate or nicotinic acid), an inverse change between
VLDL and LDL may be detected (16). In some subjects
significant increases of LDL cholesterol may occur
following CHO restriction. Carlson et al. (17) proposed
mathematical formulas, to predict changes of LDL cholesterol following diets, or drugs of the clofibrate type.

      A definitive explanation for these findings, rather
disturbing, in view of the higher atherogenicity of LDL
as compared to VLDL, is yet to be found. According to
Eisenberg (18), three possibilities should be considered:
a) increased synthesis of apo B and secretion of a larger
number of VLDL particles; b) decreased rate of removal of
intermediate lipoprotein particles from the circulation;
and c) decreased catabolism of LDL.

      Although none of these three hypothetical mechanisms
has been analysed in detail, in a recent study by Vessby
and Lithell (19), patients with hypertriglyceridemia
(particularly of type IV), treated by CHO restriction,
could be subdivided into two groups: a) those showing
an increase of LDL cholesterol and b) those showing a
decrease, the cut-point between the two being at an
LDL-cholesterol level of approximately 150 mg/dl (Fig. 2).

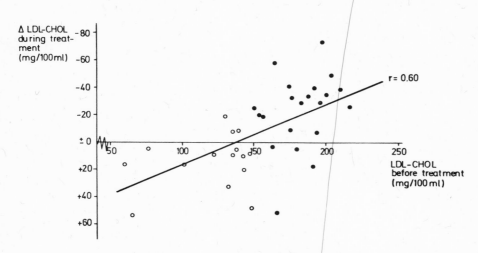

Figure 2: Variations of LDL cholesterolemia after carbohydrate restriction in type IV patients (19).

Subjects with LDL cholesterol above 150 mg/dl generally had a decrease after CHO restriction, whereas those
starting from a lower level mostly had an increase. Analysis of the lipoprotein composition of the two groups
of patients showed a different cholesterol/triglyceride
(C/TG) ratio in LDL and HDL. In particular, the C/TG
ratios were lower in patients responding to therapy with
an elevation of LDL cholesterol. These patients also had
significantly lower HDL cholesterol levels. After CHO
restriction,patients responding with increased LDL cholesterol had increased C/TG ratio in LDL, whereas no significant changes were noted in HDL composition.

According to Vessby and Lithell this data confirms
the hypothesis that there may be different subclasses of
type IV patients. Some patients have an increased number
of VLDL particles of relatively normal composition and

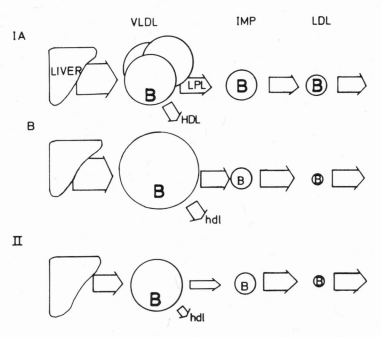

Figure 3. Possible mechanisms of VLDL catabolism in type IV
patients after carbohydrate restriction.
IA -   Normal VLDL composition and clearance (in some cases
leading to increased LDL cholesterol).
IB - VLDL of large size and decreased clearing capacity
as witnessed by low HDL.
II - Impaired removal capacity for VLDL (19).

size: these patients have an adequate clearing capacity,
sufficient to convert a normal number of VLDL particles
to LDL, every particle containing a normal amount of
apoprotein C in relation to apo B. It is possible that
some of these patients may have elevation of LDL cholesterol following the diet, if LDL clearance is impaired.
In contrast, in other patients, an accelerated VLDL synthesis, generally with increased TG content per particle may
be present; CHO restriction in these may induce decreased
TG synthesis, thus improving clearance and leading, eventually, to increased LDL cholesterol levels. This hypothesis is supported by the finding of low HDL levels,
consistent with reduced C proteins. In other words, the
final catabolic step of TG rich lipoproteins, i.e. LDL
removal, may become saturated, both when VLDL catabolism
is enhanced and, paradoxically, even when VLDL triglyceride
synthesis is reduced (Fig. 3).

EFFECT OF DIETARY FATTY ACIDS ON LIPOPROTEIN COMPOSITION

     Changes of dietary fatty acids, particularly increa-
ses of the polyunsaturated/saturated (P/S) ratio, are
widely used treatments for hyperlipidemias and expecially
for hypercholesterolemias (15). The mechanism of the
hypocholesterolemic effect of polyunsaturated (PU) fatty
acids is still the object of dispute. Studies on the
composition and structure of lipoproteins following diets
with different fatty acid compositions, both in man and
in experimental animals, have however provided findings
which may help explain the cholesterol lowering and possibility anti-atherosclerotic effects of PU fatty acids.

     Administration of diet rich in PU fatty acids (i.e.,
daily consumption of a large dose of a vegetable oil) may
lead to rapid changes in the composition of plasma fatty
acids. A doubling of the percentage of linoleate may be
observed after addition of 100 ml of sunflower oil to the
diet for just one day (20). Long-term administration of
diets rich in linoleate may lead to marked enrichment of
this fatty acid in the plasma cholesterol ester fraction
(21), thus, according to Kingsbury (22), exerting a protective effect against atherosclerosis.

     Studies on the effect of different fatty acids on
lipoprotein secretion have been reported by Ockner et al.
(23). These Authors examined the composition and distribution of lymph lipoproteins following duodenal instillation in the rat of various long chain fatty acids in the

form of mixed fatty acid-monoolein-taurocholate micelles.
All fatty acids (saturated and unsaturated) cause significant increases of chylomicron triglycerides. However,
a significant difference may be observed between the
effect of palmitate-monoolein micelles and that of
oleate-monoolein or linoleate-monoolein micelles. Whereas,
in fact, after palmitate lymph cholesterol is equally
distributed between chylomicrons and VLDL (Table 3); after
oleate and linoleate, the largest portion of lymph cholesterol is found in chylomicrons and an approximately equal
percentage in VLDL and d > 1.006 lipoproteins. Administration of progressively larger amounts of linoleate-monoolein
micelles, in particular, results in a linear increase
of chylomicron lipids and no significant changes of VLDL,
whereas after palmitate-monoolein micelles both chylomicron and VLDL fractions are lenearly increased. These
observations are of particular interest, since the half-
-life of chylomicrons in rats is approximately 10 minutes,
versus 21 minutes for VLDL, thus suggesting an easier
clearing of triglyceride-rich particles after monounsatu-
rated and polyunsaturated fatty acid feeding. These findings suggest changes in the physico-chemical properties
of lipoproteins when the diet is enriched in unsaturated
fatty acids. According to the model described by Spritz
and Mischkel (24), saturated and unsaturated fatty acids
differ in molecular configuration and in transition point
and are possibly absorbed over a different length of
intestine. Oleic and linoleic acid, absorbed over a
shorter intestinal segment,would produce a smaller number
of large transport particles, whereas palmitate, absorbed
over a longer segment of intestine, would develop smaller
particles (23).

     Studies in rabbits, fed a cholesterol rich diet, and
the same diet with the addition of corn oil (a PU rich oil)
or coconut oil (rich in saturated fat), were carried out
by Stange et al. (25). These authors noted significant
reduction of atheromatosis in the corn oil fed grup, in
spite of insignificant changes of plasma cholesterol
levels. HDL cholesterol was increased in the corn oil
treated group, as compared to the others, and interesting
differences in the electron-microscopical features were
reported. In particular, in the cholesterol and coconut
oil-cholesterol fed groups, there was frequent stacking
and fusion of the lipoprotein particles, with formation
of rosette-like structures. These were not detected in
the control and cholesterol-corn oil fed groups, where
particles were of regular size and with rare fusions.
Similar findings have been reported by Rodriguez et al.
(26), who noted, after cholesterol feeding, an enrichment

Table III

CHOLESTEROL DISTRIBUTION IN VLDL AND CHYLOMICRONS AFTER
DUODENAL INSTILLATION OF DIFFERENT FATTY ACID EMULSIONS
(23)

Duodenal	CHOLESTEROL DISTRIBUTION		
Infusion	Chylos	VLDL	d > 1.006
NaCl	16.8 %	53.9	29.3
16.0	40.3	40.3	19.5
18.1	54.9	25.9	20.0
18.2	53.7	25.5	20.9

Half-life of VLDL : 21.1 + 7.4 min
  "   "   "   Chylos: 10.3 + 2.6 "

of 18:1 in the VLDL of cholesterol-treated rabbits (VLDL
is the major form of cholesterol transport in these animals). The rosette-like formations and stacks were correlated by these Authors to the high content of sphingomyelin in cholesterol-rich lipoproteins (27).

A triglyceride lowering effect has been also described for diets rich in PU fatty acids. Chait et al. (28)
noted a triglyceride decrease of 35% when a high P/S diet
is substituted to a common saturated fat diet. This effect
is found both in normal subjects and in patients with hypertriglyceridemia. Triglyceride reduction occurs mainly in
the VLDL fraction; during the saturated fat diet the tri-
glyceride/cholesterol ratio in VLDL is 2.7, whereas after
PU fat it decreases to 2.17. During the PU diet there is
probably a reduction of VLDL synthesis. This is shown
by the lower (about 60%) incorporation of labelled linoleate in VLDL-TG as compared to labelled palmitate. This
data, partially confirming the conclusions of the study
by Ockner et al. (23) in rats, supports the concept that
there is no necessity to use different diets for hypertriglyceridemic and hypercholesterolemic patients. In
fact, just one single diet, with a high P/S ratio, may
be useful in both syndromes.

Physico-chemical properties of lipoproteins following
diets with high a P/S ratio have not been the object of

detailed investigations. Some data are reported in the studies by Peeters and Blaton on the effect of polyenyl phospholipids, presented in another paper in this course. Preliminary findings by Taunton et al. (29) have, however, indicated that diets with a very high P/S ratio significantly increase fluidity of lipoproteins (as measured by the motional properties of spin label TEMPO and by EPR of fatty acids). Lipid fluidity is insensitive to temperature and is normally maximal in VLDL as compared to LDL and HDL. After a diet with a high P/S ratio, fluidity of VLDL and LDL is significantly increased, whereas that of HDL is unmodified. The effects of dietary PU fatty acids on lipoprotein metabolism may, therefore, be due in part to increased fluidity of lipoprotein lipids.

DIETS AFFECTING PLASMA CHOLESTEROL LEVELS

Experimental induction of hypercholesterolemia

Dietary induction of hypercholesterolemia has been described very early in the study of atherosclerosis, and has provided some of the best models for research on atherogenesis and for evaluation of pharmacological, surgical and dietary means to reduce atheromatosis. Lipoprotein changes in some of the models which have received the most attention will be described here.

<u>Rabbit</u> – The earliest description of dietary induction of hypercholesterolemia in animals has been that of the cholesterol fed rabbit (30). Cholesterol feeding (1-2 g/day) rapidly induces a dramatic hypercholesterolemia in the animal, with the development of atheromatous lesions within a few weeks of treatment. Some variability of different rabbit strains in the atheromatous response has been reported (31).

Analysis of plasma lipoproteins following cholesterol induction in rabbits has shown an enrichment of cholesterol esters in lipoproteins of density less than 1.019 (VLDL and IDL) (Table 4). Cholesterol esters, normally about 5% of d < 1.019 lipoproteins, become 40-60% of these lipoproteins (26). Change in cholesterol distribution is gradual with time, as described by Brattsand (32). This Author suggests that administration of semisynthetic diets containing coconut oil or butter and a relatively low percentage of cholesterol, may induce (at plasma cholesterol concentrations of 250-400 mg/dl) a hyper-LDL cholesterolemia, similar to type II disease in man. HDL lipoproteins, markedly decreased by diets of this type, become

Table IV

LIPOPROTEIN DISTRIBUTION IN CONTROL AND HYPERCHOLESTEREMIC
RABBITS

(mg/dl; $\overline{X}$ ± SD of six rabbits)

	CONTROLS	HYPER CHOL.
VLDL	0.4 ± 0.1 (15.3%)	19.7 ± 1.5 (46.3)
IDL	0.4 ± 0.1 (15.6)	13.2 ± 1.0 (30.9)
LDL	0.7 ± 0.1 (26.8)	6.5 ± 0.1 (15.2)
HDL	1.1 ± 0.2 (42.3)	3.2 ± 0.1 (7.5)

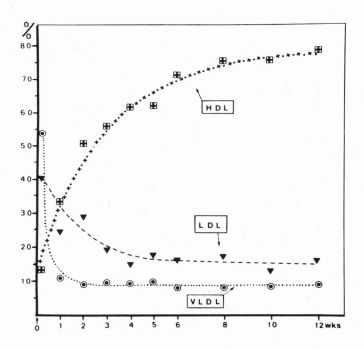

Figure 4:   Progressive free cholesterol enrichment of HDL
from cholesterol fed rabbits (33).

progressively enriched with free cholesterol (33) (Fig.4),
possibly due to increased available substrate, without
concomitant increase of lecithin cholesterol esterifying
activity.

   Lipoprotein modifications following cholesterol induction in rabbits are not restricted to the lipid composition, but include significant changes in the apoprotein moieties, particularly of d < 1.019 fractions.
Shore et al. (34) indicated in particular, a marked increase of an apoprotein component, tipical of human type III
hyperlipoproteinemia. This apoprotein, rich in arginine
("arg rich" or apo E) has been detected, as reported below,
in several cholesterol-rich lipoprotein fractions, associated to the atherosclerotic process. D<1.019 lipoproteins from cholesterol fed rabbits display significant
adhesiveness and easily show stacks, when examined by
electron microscopy (35). These ultrastructural changes,
as indicated above, may be reduced by corn-oil feeding
(25).

   Turnover experiments by Rodriguez et al. (36) on
d<1.019 lipoproteins from control and cholesterol fed
rabbits, demonstrated significantly reduced fractional
catabolic rate (FCR) of ^{125}I labelled lipoproteins from
cholesterolemic animals, and significantly increased uptake in the aortic wall, when these are injected into
control animals. Analysis of radioactivity distribution,
shows a markedly reduced conversion to lipoproteins of
higher density. These findings are similar to those
reported by Eisenberg et al. (37) for VLDL of type III
individuals.

   Guinea Pigs are particularly responsive to cholesterol administration. After feeding a diet containing
1% cholesterol, plasma, red cell and liver cholesterol
levels are significantly elevated just within 1 week.
After 10-12 weeks, tissue lipids and cholesterol increase
several times above normal levels and a fatal hemolytic
anemia usually develops (38).

   Lipoprotein analysis following dietary induction
shows significant changes both of LDL and HDL lipoproteins (Fig. 5) (39). LDL is markedly elevated in cholesterol fed, anemic guinea pigs and may be separated into
two main fractions by gel chromatography. One fraction
is similar to normal LDL, whereas to other is very rich
in free cholesterol and appears as large transparent discs
in electron microscopy. Stacked discs are also found

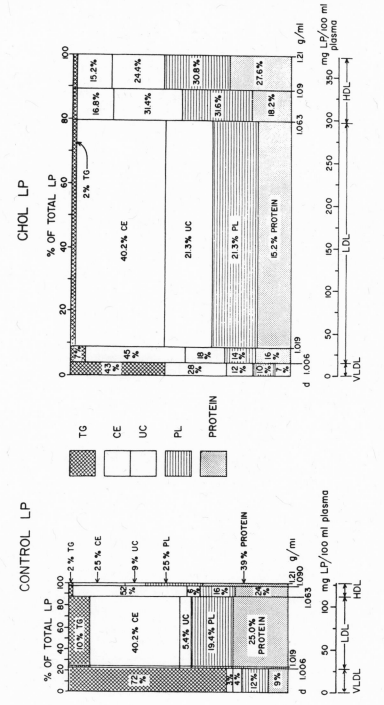

Figure 5: Lipoprotein changes following cholesterol induction in guinea pigs. A marked increase of LDL and HDL is noticeable (39).

in the HDL fraction. Apoprotein composition, both of
LDL and HDL, is suggestive of the presence of apo E-like
components, similarly to cholesterol fed rabbits.

Dogs are generally considered as refractory to atherosclerosis and poorly inducible by diet. When an atherogenic diet is however administered to thyroid-suppressed
dogs, atherosclerosis develops in a large percentage of
animals. Mahley et al. (40,41) analysed lipoprotein
composition of hypothyroid dogs fed an atherogenic diet.
These authors noted that "responders" and "non responders"
could be readily identified among animals.

Whereas "non responders" have no atherosclerosis and
a lipoprotein distribution and composition not markedly
dissimilar from control animals, "responders" show high
cholesterol concentrations in VLDL and the presence, in
these, of an apoprotein component identifiable with apo E.

Miniature Swine - Spontaneous atherosclerosis of
miniature swine can be accelerated by increased dietary
cholesterol. Addition of lard and cholesterol to the
diet of miniature swine 1-2 years of age markedly increases plasma total cholesterol levels (42). Lipoprotein
fractionation demonstrates increased cholesterol in a
lipoprotein fraction with α electrophoretic mobility and
separable at density 1.063-1.087 (43). This cholesterol-
-ester rich HDL, designated HDL-C, has a high content of
apoprotein E.

An interesting abservation related to HDL-C is that
this swine lipoprotein can inhibit HMG-CoA reductase, the
key enzyme of cholesterol biosynthesis, when incubated
with human fibroblasts (44). This finding shows that
the so called "LDL-receptor" of human fibroblasts, docu-
mentedly absent in human homozygous type II hyperlipoproteinemia (45), can interact not only with apoprotein B,
as generally believed, but also with other apoprotein
components, i.e. probably apo E. Moreover, interactions
with the "LDL-receptor" are not species specific, but may
occur also with lipoproteins from different species.

Non Human Primates are a choice animal model for
the induction of atherosclerosis, due to the documented
similarities of their atherosclerotic lesions with those
of man. Primates differ, however, remarkably in the
dietary inducibility of hyperlipidemia and atherosclerosis
(46). Some species are practically refractory to the

dietary induction of hyperlipidemia and atherosclerosis, whereas others are very sensitive. In spite of the numerous studies on the atherogenic effects of different diets in primates, contributions on the lipoprotein changes following these diets are not abundant. Of particular interest, for most investigators, has been the analysis of lipoprotein similarities between primate and human hyperlipoproteinemias (46).

Rhesus monkeys have been the most thoroughly studied species. They are very sensitive to dietary cholesterol and develop atherosclerosis after several months of treatment. By monitoring lipoprotein distribution in Rhesus monkeys subjected to high dietary cholesterol, according to the Rudel procedure (47) with agarose gel--cromatography, it was possible to show significant alterations in the distribution of lipoproteins (Fig. 6). In particular, Macaca mulatta shows elevated plasma VLDL, IDL and LDL, with decreased HDL. LDL size is increased, as estimated from elution volumes; VLDL, markedly enriched with cholesterol esters, acquire β mobility on agarose electrophoresis. Therefore, lipoproteins of cholesterolemic Macaca have a distribution intermediate between that of human type II hyperlipoproteinemia and that of hypercholesterolemic rabbits.

A different pattern is that of the squirrel monkeys (Saimiris sciureus). After cholesterol induction, the agarose gel profile shows a striking increase of VLDL. LDL and HDL are not significantly changed. VLDL, similarly to rabbits, have increased cholesterol esters and electrophoretic β mobility. In line with the lipoprotein differences, the extent and distribution of atherosclerotic lesions is different between Macaca mulatta and Saimiris (48).

In the vervet monkeys (Cercopithecus ethiops) the pattern is still different. These animals are less responsive to the dietary induction of cholesterolemia. However, their chromatography elution profile indicates a dramatic increase of LDL, with slight reduction of HDL. LDL composition is not markedly modified. These monkeys are probably the most similar, in lipoprotein distribution, to the human type II hyperlipoproteinemia. The hypercholesterolemic response, maximal at 1 mg cholesterol/Kcal/day, is in the 300-400 mg/dl range, similar to that of the human disease. Distribution and morphology of atherosclerotic lesions have also been found to bear resem-

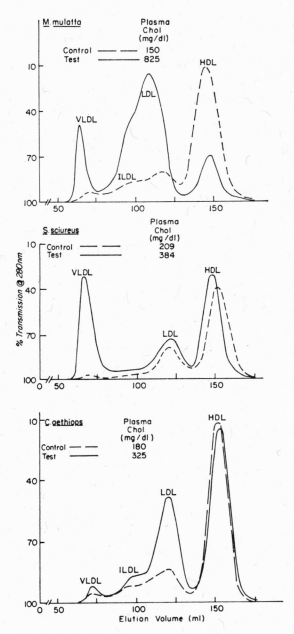

<u>Figure 6</u>: Changes of plasma cholesterol and lipoprotein distribution, analyzed by agarose gel chromatography, in primates fed atherogenic diets (46).

blance to those of human atherosclerosis (49).

Apoprotein composition of atherosclerotic monkeys
has not been studied in great detail. It is generally
assumed that LDL lipoproteins have a predominance of the
B apoprotein and that an apo E may be found in choleste-
remic VLDL. Interesting differences were however found
in the HDL apoprotein and phospholipid composition of
different primates, when compared to humans (50). It was
in fact observed that, whereas chimpanzee has a phospholipid pattern in HDL, almost indentical to that of man
(particularly rich in sphingomyelin and phosphatidylcholine,
relatively low in phosphatidylinositol and phosphatidylethanolamine), baboon and Rhesus HDL are very high in
phosphatidylcholine and relatively low in sphingomyelin
as compared to humans and chimpanzee. Following dietaty
induced hypercholesterolemia, chimpanzee, like man, has
an increased sphingomyelin/phosphatidylcholine ratio.
This is also found in Rhesus, but not in baboon. Chimpanzee has a typical hyper β-lipoproteinemia following
cholesterol (it may have a hyper-pre β lipoproteinemia
after sucrose and has a definitely decreased apo A1 in
both of these conditions) (50).

Turnover studies on primate lipoproteins have been
described by Portman et al. (51). In squirrel monkeys,
who, as pointed out, are similar to rabbits in their lipoprotein distribution following diet induced hypercholesterolemia, a slight reduction of LDL turnover is caused
by hyperlipidemia, with increased absolute rates of
synthesis and catabolism. In another study (52) the same
Authors compared the effects of different dietary fats
added to the cholesterol rich regimen. A higher fractional turnover rate of LDL was noted in monkeys when safflower oil was added to diets rich in cholesterol alone, butter + cholesterol, or butter alone. The fraction of the
LDL pool remaining intravascular is also lower on diets
containing safflower oil.

Avian Models- Birds, in particular pigeons and quails,
respond with an elevation of plasma cholesterol levels
following appropriate diets, cholesterol being carried
predominantly in LDL (46).

They also display significant atherosclerotic lesions.
The Japanese quail (Coturnix japonica) has elevated
plasma cholesterol following a cholesterol-cholic acid
rich diet; it also shows significant incidence of

severe aortic and brachio-cephalic atherosclerosis.
Chapman et al. (53) have described a selective breeding
method for obtaining highly responsive quails.

Among pigeons, the White Carneau and Show Race breeds
have a significantly different inducibility of atherosclerosis by diet (54). These two species differ in the oleic-
-linoleic acid ratio in aortic sterol esters. Comparison
of responder and non responder pigeons has failed to describe specific differences in the cholesterol metabolism
of these animals (55).

Rats - In spite of their extensive use for physiological studies on lipoprotein metabolism, rats have not been
examined in detail for specific lipoprotein changes following atherogenic diets. Following addition of cholesterol to the diet, d < 1.006 lipoproteins are increased,
and a new lipoprotein (d 1.006-1.030) may be detected,
which contains immunologically identifiable HDL and LDL
(56).

On cholesterol-cholic acid rich diets, added with
propyl thiouracil, rats develop significant hypercholesterolemia and some atherosclerotic lesions. Breslow et
al.(57) recently indicated that lipoproteins of d < 1.063,
isolated from these rats, significantly suppress HMG-CoA
reductase in rat liver cell cultures. This effect is not
noted with lipoproteins from control animals.

Clinical Studies

In humans, addition of cholesterol to the diet, results in increased plasma cholesterol levels, predictable, in the general population, according to the formula

$$\Delta \text{ Chol} = 1.5 \sqrt{\Delta \text{ Chol}/1,000 \text{ KCal}}$$

suggested by Keys et al. (58). Analysis of individual
responses shows, however,a considerable variability.
Quintao et al. (59) in particular, showed that addition
of cholesterol to the diet beyond certain limits fails
to markedly increase plasma cholesterol levels. These
Authors suggested that administration of very large
cholesterol doses does not allow adequate absorption and,
moreover, an extention of extra-vascular cholesterol pools
may be at hand.

Recently, Nestel and Poyser (60) examined in detail

plasma cholesterol changes following two different daily
intakes of cholesterol (250 and 750 mg). They observed
that, of the eight tested patients, only three showed
marked increases of cholesterolemia (all three had
hypercholesterolemia), the other five having practically
no change. On the whole, the mean cholesterol change
of the total group was consistent with Keys' formula.
Analysis of cholesterol absorption and excretion (both
of acidic and neutral metabolites), indicated that, in
the majority of patients, increased cholesterol intake
is fully compensated by increased excretion, mostly
under the form of a neutral sterols, and decreased synthesis. In some patients, cholesterol may be however retained, with a lesser change of neutral sterol excretion and
little change of synthesis. According to the Authors,
this data suggests that in some patients the cholesterol
transport form may be changed to one less readily degraded.

Lewis (personal communication) has provided evidence
that exogenously administered cholesterol may be transported in a VLDL fraction, thus giving rise to a form of
type III hyperlipoproteinemia, similarly to the rabbit
or to other animal species, as above outlined. This data
awaits further confirmation, and may partly explain discrepancies in studies related to the dietary induction of
hypercholesterolemia in man.

REDUCTION OF HYPERCHOLESTEROLEMIA

Diet induced hypercholesterolemia in animals may be
readily reversed by changing the animal diet to the original type, thus achieving, at time, a regression of atherosclerotic lesions (61). In man, on the contrary, hypercholesterolemia is often of genetic origin and may be observed also on a "normal" diet, not particularly rich in
cholesterol or saturated fat. Dietary treatment includes,
therefore, reduction of dietary cholesterol and increased
P/S ratio. Lipoprotein changes following increased P/S
ratio have been reported in a preceding part of this paper. It is generally agreed that a high P/S ratio will
cause a significant reduction of plasma cholesterol and
particularly of the LDL fraction (62).

Classic dietary treatments reducing plasma cholesterol
levels do not in general markedly affect lipoprotein composition. This has been well documented by Bagnall (63),
who noted that in children with familial hypercholesterolemia, reduction of serum cholesterol fails to modify

LDL composition (Table 5). It may be assumed that cholesterol reducing diets will cause a drain of cholesterol,
mostly from LDL and VLDL, not significantly modifying
lipoprotein composition. Aside from the classic dietary
treatments, new approaches have been recently described
for the more severe forms of hypercholesterolemia. One
of these is the intravenous hyperalimentation (64). This
type of treatment, i.e. administration i.v. of solutions
containing high percentages of dextrose and nitrogen,
strikingly decreases plasma cholesterol levels also in
patients with the homozygous form of type II hyperlipidemia. This approach is derived from the treatment of
glycogen-storage disease, where hyperlipidemia is also
reduced in this way, and which may find a definitive cure
by portocaval shunt. I.v. hyperalimentation is now used
as a presumptive test on the potential cholesterol lowering
effect of portocaval shunt.

Following i.v. hyperalimentation marked decreases
of plasma total and LDL cholesterol levels are noted.
HDL cholesterol also decreases dramatically (Table 6).
The relative composition of LDL after therapy is characterized only by a slight increase of triglycerides. LDL
turnover in these patients is not significantly changed.
The discussion on whether the mode of action of i.v.
hyperalimentation is similar to that of portocaval shunt,
has not brought to definitive conclusions. It is likely
that the mechanism of i.v. hyperalimentation is more
complex than, as suggested for portocaval shunt, elimination of the direct exposure of the liver to pancreatic
hormones (65). The decrease of HDL is very significant
and also hard to explain, but it is probably related to
the mechanism of action of this special dietary treatment
for hypercholesterolemia.

A simpler dietary approach for hypercholesterolemia
has been recently suggested by Sirtori et al. (66), by
substituting animal proteins in the diet with a soybean
textured protein. Possibly due to aminoacid differences
in the animal and vegetable proteins (67), plasma cholesterol levels are markedly decreased within two-three
weeks in patients treated with the soybean regimen.
Analysis of lipoprotein changes following this diet have
shown that both LDL and VLDL cholesterol are significantly
decreased (Table 7), whereas HDL cholesterol is practically unchanged. This data is related to 20 patients treated
with soybean together with a diet with a high P/S ratio.
Preliminary data on patients treated with the same regimen and low P/S ratio has shown that reduction of LDL

Table V

EFFECT OF DIET ON LDL COMPOSITION IN CHILDREN WITH FAMIL-
IAL HYPERCHOLESTEROLEMIA (63)

	Normal	Type II	Diet
Chol.	197 + 27	367 + 55	265 + 43

% Composition of LDL

	Normal	Type II	Diet
CE	42.7 + 2.7	45.2 + 2.0[†]	45.1 + 1.7[†]
FC	11.4 + 1.4	12.7 + 0.9[†]	12.2 + 1.8
PL	21.5 + 2.0	22.4 + 2.3	23.3 + 1.4
TG	6.0 + 2.3	3.3 + 1.0[††]	3.3 + 1.9[††]
P	18.7 + 2.1	16.4 + 1.8	16.0 + 1.9

[†] P < 0.05; [††] P < 0.01 from Normals

Table VI

LIPID AND LIPOPROTEIN RESPONSE TO I.V. HYPER-
ALIMENTATION (64)

		Total Chol.	mg/dl	LDL	Apo
		pre	post	pre	post
	1	506	412	249	201
Patients	2	545	409	272	207
	3	594	333	299	163

		HDL	Chol.	"	Triglycerides	
		pre	post		pre	post
	1	21	11		61	70
Patients	2	26	10		85	120
	3	26	17		133	115

Table VII

LIPOPROTEIN CHOLESTEROL CHANGES FOLLOWING THE SOYBEAN
PROTEIN DIET (66)[*]

20 type II patients   3 weeks of soybean

CHOLESTEROL   mg/dl

		Pre	Post	Δ
$\overline{X}$	VLDL	52	31	21[+]
$\overline{X}$	LDL	233	186	47[++]
$\overline{X}$	HDL	40	39	1
	TOTAL	325	256	69[++]
TRIGLYCERIDES		197	165	32[+]

[+]$P < 0.05$; [++]$P < 0.01$

[*]mean levels of 20 patients treated with the soybean
protein diet for three weeks, either before or after
a low-lipid low-cholesterol diet.

cholesterol is less significant in these, whereas total
cholesterolemia is decreased not differently than with
a high P/S ratio. More detailed studies on lipoprotein
composition in these patients are being carried out.

CONCLUSIONS

Dietary changes may modify lipoprotein composition,
in a few cases dramatically. Examples of these are
the animal models of hypercholesterolemia, of ethanol
or CHO induced hypertriglyceridemias in humans, and of
intravenous hyperalimentation in hypercholesterolemic
patients. In other cases, modifications may be slight
or hardly significant. Detailed compositional studies
of plasma lipoproteins are awaited to exactly determine
site and mode of action of the different diets.

REFERENCES

1. Quarfordt, S.H., Frank, A., Shames, D.M., Berman, M. and D. Steinberg. 1970. Very low density lipoprotein triglyceride transport in Type IV hyperlipoproteinemia and the effects of carbohydrate-rich diets. J. Clin. Invest 49: 2281-2297

2. Wilson, D.E., Schreibman, P.H., Breston, A.C., and R.A. Arky. 1970. The enhancement of alimentary lipemia by ethanol in man. J. Lab. Clin. Med. 75: 264-273.

3. Schonfeld, Y.1970. Changes in the composition of very low density lipoprotein during carbohydrate induction in man. J. Lab. Clin. Med. 75: 206-211.

4. Ruderman, N., Jones, A.L., Krauss, R., and E. Shafrir. 1971. A biochemical and morphologic study of very low density lipoproteins in carbohydrate induced hypertriglyceridemia. J. Clin. Invest. 50: 1355-1368

5. Carlson, L.A., and D. Ballantyne. 1976. Changing relative proportions of apolipoproteins CII and CIII of very low density lipoproteins in hypertriglyceridemia. Atherosclerosis 23: 563-568.

6. Havel, R.J., Fielding, C.J., Olivecrona, T., Shore V.G., Fielding P.E., and T. Egelrud. 1973. Cofactor activity of protein components of human very low density lipoproteins with hydrolysis of triglycerides by lipoprotein lipase from different sources. Biochemistry 12: 1828-1832

7. Schonfeld, G., Weidman, S.W., Witztum, J.L., and R.M. Bowen. 1976. Alterations in levels and interrelations of plasma apolipoproteins induced by diet. Metabolism 25: 261-275.

8. Mendelson, J.H., and N.K. Mello. 1973. Alcohol-induced hyperlipidemia and beta lipoproteins. Science 180: 1372-1374

9. Puglisi, L. Caruso, V., Conti, F., Fumagalli, R., and C. Sirtori. 1977. Effect of hypolipidemic and hypoglycemic drugs on ethanol induced hypertriglyceridemia in rats. Pharm. Res. Comm. 9: 71-77

10. Losowsky, W.S., Jones, D.P., Davidson, C.S., and C.S. Lieber. 1963. Studies of alcoholic hyperlipidemia and its mechanism. Am. J. Med. 35: 794-803.

11. Avogaro, P., and G. Cazzolato. 1975. Changes in the composition physico-chemical characteristics of serum

lipoproteins during ethanol-induced lipaemia in
alcoholic subjects.   Metabolism 24: 1231-1242.

12. Avogaro P. 1976. Ethanol-lipaemia , a doubtful risk
    factor. Proc. IV Int. Symposium on Atherosclerosis.
    Tokyo, in press.

13. Sirtori, C.R., Biasi, G., Vercellio, G., Agradi, E.,
    and E. Malan. 1975. Diet, lipids and lipoproteins in
    patients with peripheral vascular disease. Am. J.
    Med. Sci. 268: 325-332.

14. Miller, G.J. and N.E. Miller. 1975. Plasma high-den-
    sity lipoprotein concentration and development of
    ischaemic heart - disease. Lancet i: 16-19.

15. Levy, R.I., Bonnell, M., and N.D. Ernst. 1971. Die-
    tary management of hyperlipoproteinemia. J. Am. Ass.
    58:406-416.

16. Wilson D.E. and R.S. Lees. 1972. Metabolic relation-
    ships among the plasma lipoproteins. J. Clin. Invest.
    51: 1051-1057.

17. Carlson, L.A., Olsson, A.G., Orö, L., Rössner, S.,
    and G. Walldius. 1974. Effects of hypolipidemic
    regimens on serum lipoproteins. Proc. III Int. Sympo-
    sium on Atherosclerosis, Berlin, 768-781.

18. Eisenberg, S. 1976. Lipoprotein metabolism and hyper-
    lipemia. In: Atherosclerosis Reviews Vol. 1 Paoletti
    R., Gotto A.M., Eds. Raven Press. N.Y.,23-60.

19. Vessby, B., and M. Lithell.1976. Dietary effects on
    lipoprotein levels in hyperlipoproteinemia. Delinea-
    tion of two subgroups of endogenous hypertriglyceri-
    demia. Artery 1: 63-85

20. Belin, J., Smith, A.D., and R.H.S. Thompson. 1975.
    Effect of short-term oral administration of sunflower
    seed oil on the pattern of non-esterified fatty acids
    in human plasma. Clin. Chim. Acta 61: 95-98.

21. Kingsbury, K.J., Brett, C., Stovold, R., Chapman, A.,
    Anderson, J., and D.M. Morgan. 1974. Abnormal fatty
    acid composition of human atherosclerosis. Postgrad.
    Med. J. 50: 425-440

22. Kingsbury, K.J., Morgan, D.M., Stovold, R., Brett,
    G.G., and J. Anderson. 1969. Polyunsaturated fatty
    acids and myocardial infarction. Lancet ii: 1325-
    -1331.

23. Ockner, R.K., Hughes, F.B. and K.J., Isselbacher.
    1969. Very low density lipoproteins in intestinal

lymph: role in triglyceride and cholesterol transport during fat absorption. J. Clin. Invest. 48: 2367-2373.

24. Spritz, N. and M.A. Mischkel. 1969. Effects of dietary fats on plasma lipids and lipoproteins: an hypothesis for the lipid-lowering effect of unsaturated fatty acids. J. Clin. Invest. 48: 78-86.

25. Stange, E., Agostini, B., and J. Papenberg. 1975. Changes in rabbit lipoprotein properties by dietary cholesterol, and saturated and polyunsaturated fats. Atherosclerosis 22: 125-148.

26. Rodriguez, J.L., Catapano, A., Ghiselli, G.C., and C.R. Sirtori. 1976. Very low density lipoproteins in normal and cholesterol-fed rabbits: lipid and protein composition and metabolism. Part. 1. Chemical composition of very low density lipoproteins in rabbits. Atherosclerosis 23: 73-83.

27. Jackson, R.L. and A.M. Gotto jr. 1974. Phospholipids in biology and medicine (second of two parts). N. Engl. J. Med. 290: 87-93.

28. Chait, A., Onitiri, A., Nicoll, A., Rabaya, E., Davies, J., and B. Lewis. 1974. Reduction of serum triglyceride levels by polyunsaturated fat. Studies on the mode of action and on very low density lipoprotein composition. Atherosclerosis 20: 347-364.

29. Taunton, O.D., Morrisett, J.D., Segura, R., Pownall, M.J., Jackson, R.L., and A.M. Gotto. 1974. Effect of dietary fatty acid composition on lipoprotein structure. Circulation 49-50, Suppl. III, Abs. 1050

30. Anitschkow, N.N. 1913. Über die Veränderung der Kaninchen aorta bei experimenteller Cholesterinsteatose. Beitr. path. Anat. 56: 379-383.

31. Shore, B., and V. Shore. 1976. Rabbits as a model for the study of hyperlipoproteinemia and atherosclerosis. Adv. Exp. Med. Biol. 67:123-141.

32. Brattsand, R. 1976. Distribution of cholesterol and triglycerides among lipoprotein fractions in fat-fed rabbits at different levels of serum cholesterol. Atherosclerosis 23: 97-110, 1976.

33. Pinon, J.C., and A.M. Bridoux. 1977. High density lipoproteins in cholesterol fed-rabbits: progressive enrichment with free cholesterol. Artery 3: 59-71.

34. Shore, V.G., Shore, B., and R.G. Hart. 1974. Changes in apolipoproteins and properties of rabbit very low density lipoproteins on induction of cholesterolemia.

Biochemistry 13: 1579-1582.

35. Agostini, B., Seidel, D. and M. Wieland. 1973. Electron microscopy of low-density plasma lipoproteins in patients with Type IIa hyperlipoproteinemia. Naturwissenschaften 60: 111-113.

36. Rodriguez, J.L., Catapano, A., Ghiselli, G.C., and C. Sirtori. 1976. Very low density lipoproteins in normal and cholesterol-fed rabbits: lipid and protein composition and metabolism. Part 2. Metabolism of very low density lipoproteins in rabbits. Atherosclerosis 23:85-96

37. Bilheimer, D.W., Eisenberg, S. and R.I. Levy. 1971. Abnormal metabolism of very low density lipoproteins (VLDL) in Type III hyperlipoproteinemia. Circulation 44: Suppl. II, Abs. 186.

38. Green, M.H., Crim, M., Traber, M., and R. Ostwald. 1976. Cholesterol turnover and tissue distribution in the Guinea Pig in response to dietary cholesterol. Nutrition 106: 516-528.

39. Sardet, C., Hansma, H., and R. Ostwald. 1972. Characterization of guinea pig plasma lipoproteins: the appearance of new lipoproteins in response to dietary cholesterol. J. Lipid. Res. 13:624-639.

40. Mahley, R.W., and K.H. Weisgraber. 1974. Canine lipoproteins and atherosclerosis I. Isolation and characterization of plasma lipoproteins from control dog. Circ. Res. 35: 713-721.

41. Mahley, R.W., Weisgraber, K.H., and I. Innerarity. 1974. Canine lipoproteins and atherosclerosis II. Characterization of the plasma lipoproteins associated with atherogenic and non-atherogenic hyperlipidemia. Circ. Res. 35: 722-733.

42. Florentin, R.A., and S.C Nam. 1968. Dietary-induced atherosclerosis in miniature swine. I. Gross and light microscopy observation: Time of development and morphologic characteristics of lesions. Exp. Mol. Pathol. 8: 263-301.

43. Mahley, W.R., Weisgraber, K.H., Innerarity, T., Brever, H.B., and G. Assman. 1975. Swine lipoproteins and atherosclerosis. Changes in the plasma lipoproteins and apoproteins induced by cholesterol feeding. Biochemistry 14: 2817-2823.

44. Bersot, T.P., Mahley, R.W., Brown, M.S., and J.L. Goldstein. 1976. Interaction of swine lipoproteins

with the low density lipoprotein receptor in human fibroblasts. J. Biol. Chem. 251: 2395-2398.

45. Goldstein, J.L., and M.S. Brown. 1973. Familial hyper-cholesterolemia: identification of a defect in the regulation of 3-hydroxy-3-methylglutaryl-coenzime A reductase activity associated with overproduction of cholesterol. Proc. Nat. Acad. Sci. 70: 2804-2808

46. Clarkson, T.B., Prichard, R.W., Bullock, B.C., St. Clair, R.W., Lehner, N.D.M., Jones, D.C., Wagner, W.D., and L.L. Rudel. 1976. Pathogenesis of athero-sclerosis: some advances from using animal models. Exp. Mol. Path. 24:264-286.

47. Rudel, L.L., Lee, J.A., Morris, M.D., and J.M. Felts. 1974. Characterization of plasma lipoproteins separated and purified by agarose-column chromatography. Biochem. J. 139: 89-95.

48. Clarkson, T.B., Prichard, R.W., Bullock, B.C., Lehner, N.D.M., Lofland, H.B., and R.W. St. Clair. 1970. Animal models of atherosclerosis. Animal models. Biomed. Res. 3: 22-41.

49. Bullock, B.C., Lehner, N.D.M., Clarkson, T.B., Feldner, M.A., Wagner, W.D., and H.B. Lofland. 1975. Comparati-ve primate atherosclerosis I: Tissue cholesterol concen-tration and pathologic anatomy. Exp. Mol. Pathol. 22: 151-175.

50. Blaton, V., and H. Peeters. 1976. The non-human pri-mates as models for studying human atherosclerosis: studies on the Chimpanzee, the Baboon and the Rhesus Macacus. Adv. Exp. Med. Biol. 67:33-64.

51. Portman, O., Illingworth, D.R., and M. Alexander. 1975. The effects of hyperlipidemia on lipoprotein metabolism in Squirrel monkeys and rabbits. Biochim. Biophys. Acta. 398: 55-71.

52. Portman, O., Alexander, M., Tanaka, N., and P. Soltys. 1976. The effects of dietary fat and cholesterol in the metabolims of plasma low density lipoprotein apo-proteins in Squirrel monkeys. Biochim. Biophys. Acta. 450:185-196.

53. Chapman, K.P., Stafford, W.W., and C.E. Day. 1976. Animal model for experimental atherosclerosis produced by selective breeding of japanese quail. Adv. Exp. Med. Biol. 67:347-356.

54. Ravi Subbiah, M.T., Kottke, B.A., and I.A. Carlo. 1974. Studies in spontaneously atherosclerosis

susceptible and resistant pigeons: nature of plasma and aortic sterol, steryl esters, and free fatty acids. Int. J. Biochem. 5, 63-68.

55. Wagner, W.D., and T. B. Clarkson. 1974. Mechanisms of the genetic control of plasma cholesterol in selected lines of Show Racer pigeons. Proc. Soc. Exp. Biol. Med. 145:1050-1057.

56. Lasser, N.L., Roheim, P.S., Edelstein, D. and H.A. Eder. 1973. Serum lipoproteins of normal and cholesterol-fed rats. J. Lipid. Res. 14: 1-8.

57. Breslow, J.L., Spaulding, D.R., Lothrop, D.A., and A. Clowes. 1975. Effect of lipoprotein on 3-hydroxy--3-methylglutaryl coenzyme A (HMG CoA) reductase activity in rat liver cell culture: special suppressant effect of a lipoprotein isolated from hypercholesterolemic rat plasma. Biochem. Biophys. Acta. 67: 119-125.

58. Keys, A., Anderson, J.T., and F. Grande. 1965. Serum cholesterol response to changes in the diet II. The effect of cholesterol in the diet. Metabolism 14: 759-765.

59. Quintao, E., Grundy, S.M., and E.H. Ahrens jr. 1971. Effects of dietary cholesterol on the regulation of total body cholesterol in man. J. Lipid. Res. 12: 233-247.

60. Nestel, P.J., and A. Poyser. 1976. Changes in cholesterol synthesis and excretion when cholesterol intake is increased. Metabolism 25: 1591-1599.

61. Armstrong, M.L., 1976. Regression of atherosclerosis. In: Atherosclerosis Reviews Vol. 1. Paoletti R., Gotto A.M., Eds. Raven Press. N.Y., 137-182.

62. Levy, R.I. 1972. Dietary and drug treatment of primary hyperlipoproteinemia. Ann. Int. Med. 77: 267-294.

63. Bagnall, T.F. 1972. Composition of low density lipoprotein in children with familial hyperbetalipoproteinemia and the effect of treatment. Clin. Chim. Acta. 42:229-233.

64. Torsoik, H., Feldman, H.A., Fischer, J.E., and R.S. Lees. 1975. Effects of intravenous hyperalimentation on plasma-lipoproteins in severe familial hypercholesterolaemia. Lancet i: 601-607.

65. Ahrens, E.H. jr. 1974. Homozygous hypercholesterolemia and the portacaval shunt. Lancet. ii, 449-451.

66. Sirtori, C.R., Agradi, E., Conti, F., Mantero, O.,

and E. Gatti. 1977. Soybean-protein diet in the treatment of type II hyperlipoproteinemia. Lancet $\underline{i}$: 275-278.

67. Carroll, K.K., and R.M.G. Hamilton. 1975. Effects of dietary protein and carbohydrate on plasma cholesterol level in relation to atherosclerosis. J. Food Sci. $\underline{40}$: 18-23.

Conclusion

LIPOPROTEIN MOLECULE OR LIPOPROTEIN PARTICLE?

Hubert Peeters

Lipid and Protein Department, LBS

B-1180 Brussels, Belgium

May I wind up this Advanced Study Institute by sketching briefly what this course has meant at least as far as my judgement goes.
Not that I believe to be a polyvalent expert on all topics presented
but I do think it is my duty to outline the status of our knowledge
and the avenues along which we should make progress. What do we expect to happen, what is to become important, what should be looked
for?

The first line of thought is the apoprotein line. This line
has been complex but rewarding and has been shedding light on a set
of clinical problems nobody hoped to solve a few years ago. The
beautiful thing about apoproteins is their potential in the inter-
pregation of the ultracentrifugal and electrophoretic protein patterns. Two physical parameters - gravitational density during centrifugation and optical density after electrophoresis - were in need
of an objectivation through the primary structure of the apopolypeptides. The distribution of the apopolypeptides interprets quite a
few problems of lipoprotein density classes but does not solve the
entire problem. If only the exchange of lipids during centrifugation
under the effect of high salt concentrations does call for further
careful and critical assessment of UCF data. In my view the recentri-
fugation of electrophoretic lipoprotein fractions is also a must.
With modern knowledge on apoproteins and their quantitation, completed
by a careful and complete lipid analysis such as of the phospholipids,
a profound and motivated comparative countercheck of ultracentrifugal
and electrophoretic fractions should be performed. Several times in
the course of these days important physiological shifts inside density classes were documented. This confirms my opinion that still
more comparative work is required.

Systematics in lipoproteins will continue to fulfill a unique role: they absorb and integrate knowledge from an increasing number of disciplines and in turn generate important data which are useful to these other disciplines by identifying new species of molecules and by shedding light on their interrelationships.

We touched on two problems: apoprotein quantitation and lipid quantitation. We should accept that neither one of these problems is really solved. Elegant and efficient methods can be expected and are indeed needed to rapidly and precisely quantitate proteins and lipids. Where is the comprehensive paper on delipidation? What is the comprehensive task of apoprotein? Most problems with lipoproteins originate from their mysterious amphiphilic properties. All our analytical methods struggle with two antithetic requirements: to isolate valid protein and valid lipid out of a molecule which has been built to swim like a fish in water. A dry duck is a lame duck: an apoprotein seems hard to dissolve. On the contrary a lipoprotein is beautifully stabile in our aqueous biological fluids.

As a consequence of this problem the importance of physical methods which can look at the intact molecule becomes evident and this is a field where more progress is to occur. In the past centrifugation and electrophoresis did more for lipoproteins than all cholesterol determinations. And cholesterol anyhow became truly interesting only when related to alpha or beta protein, thus when related to lipoprotein. Here then is an incidental remark: we need to introduce molar expression of lipids and in turn express these in function of lipoproteins: if we do not succeed in doing this the clinicians will persevere for a long time in a rather rough approximation of dyslipaemia as a kind of sophisticated form of hypercholesterolaemia whereas the core of the matter lies in a profound and disturbed phospholipid-apoprotein relationship followed or preceded by triglyceride and cholesterol ester disorders. If we group these remarks the need for estimations and parameter evaluations of intact lipoproteins is a corollary as well as a prerequisite. A method such as microviscosity may be a step in the good direction.

Which of these questions is the prime mover? All approaches to interpretation of the metabolic pathways of lipoproteins are still blocked by a set of serious questions:
- Site and rate of synthesis of apoproteins
- Mechanisms of lipidation after apoprotein synthesis
- Role of succession of incorporation of lipid classes
- Interexchanges of lipids between secreted lipoproteins
- Catabolism of lipoprotein lipids
- Relipidation and reuse of remnant apoproteins after their delipidation in vivo.

One could organize quite a few round tables around such topics.

We shall go home with such questions unanswered.

To my mind any physical method aimed at the structura interpretation of the lipoprotein molecule is due to allow for a fundamental interpretation of lipoprotein metabolism. The metabolic
problems will not be solved through metabolic methods alone, the
metabolic problems require a profound knowledge of the qualities of
the lipoprotein molecule itself. Physical parameters can be much
shorter to obtain than a panel of chemical data.

Evidently comparative biochemistry including apoprotein and
lipid analysis cannot be overlooked. Such studies are easier to
perform than the work on the exceedingly rare congenital disorders
in man. Animals will become an excellent tool when the knowledge
of the biophysical relationship between protein and lipid will be
better understood because extrapolation, for instance of data on
drug effects, will thus become feasible. This is not yet possible
today.

Is the word lipoprotein molecule correct? I guess not: we
should have discussed the lipoprotein particle and this criticism
of the title of the Institute itself, is proof of our awareness of
our ignorance. To attack lipoproteins - molecules or particles -
also requires the help of immunology. I touched on this problem
indirectly when I mentioned the quantitation problem, but immunology of lipoproteins indeed has its home-made problems: antisera
see antigens when they are naked. Do the lipids cover up antigenic
sites? How often? How much? These remarks on immunology bring us
into the heart of the matter: lipoproteins are not "beautiful
people" for a few selected scientists. They actually hurt the vessel wall and this kills people, including the beautiful. So the
natural history, the pathology, and the search for drugs and diets
are of the utmost importance. This part of the message, though
stated very briefly, should stay with us.

What do I suggest to the new students in the field?
a. To read the older papers. Several times during these days facts
   have been rediscovered which most of us had forgotten.
b. To remark that people die in spite of our perfectly equipped
   laboratories.
c. To ask for more and for better laboratories. Scramble for new
   physical and clinical methods as well for apoprotein as for
   lipid analysis. Make comparative studies so that your progress
   can be used over a broad field of research in the lipoprotein
   field to which we belong.
d. To establish contact not only with your kinsmen but with people
   of different disciplines in the field of dyslipaemia. The same
   sample can be looked at by several people.
e. To exchange methods, standards, procedures on the practical level. We have been deceptively poor in this respect, I know

this. At least bridges were built, contacts were established.

   A second set of propositions covers a set of topics in which I
believe:
1. Apoprotein quantitation as a clinical tool
2. Phospholipid distribution in lipoproteins
3. Link plasma, platelet, and arterial wall lipids, especially
   phospholipids
4. Look for acyl transfer mechanisms requiring other phospholipids
   than lecithin as a substrate
5. If you have a physical method available, become an expert, stick
   to your guns. Use precise polypeptides and serial systematic
   lipids preferably the synthetic and semi-synthetic ones. Do not
   use a physical method on a lipoprotein until you know what exactly happened with a model system.
6. Fatty acids were not mentioned till this morning. Reread then
   the old papers, reinterpret them and repeat the old clear cut
   experiments that withstood the tug of time, and look now for
   fatty acid movements inside lipoprotein classes, and inside their
   lipids.
7. Use experimental animals only if you have access to polypeptide
   and lipid data. Do not feed rabbits, SEA quail, or pigeons, rapidly measure their cholesterol and write a paper. You cannot
   correctly assess nor evaluate changes in lipids if you cannot
   relate them to apoprotein content.
8. Participate in any drug trial they propose to you but only if
   the company accepts your minimal clinical and laboratory guide-
   lines as referred to in the previous point.

   What do we expect to achieve ourselves?
- To develop SM/PC ratio as a clinical concept (sphingomyelin/leci
  thin)
- To develop microviscosity as a physical measurement of lipoprotein
- To discover carrier proteins for phospholipids. They should act
  as messengers at the site of lipoprotein synthesis. They have to
  be specific for each phospholipid and sensitive to the fatty acid
  load of the phospholipid. If you consider the chronobiology of
  cholestereol synthesis in the rat for instance, such rythmic succession of specific enzyme and carrier proteins will be requested.

   The informational potential of a lipoprotein is tremendous: if
we persevere we shall break its code for the benefit of mankind and
just a little to our own satisfaction.